SpringerBriefs in Economics

More information about this series at http://www.springer.com/series/8876

Isamu Matsukawa

Consumer Energy Conservation Behavior After Fukushima

Evidence from Field Experiments

 Springer

Isamu Matsukawa
Faculty of Economics
Musashi University
Nerima, Tokyo
Japan

ISSN 2191-5504   ISSN 2191-5512   (electronic)
SpringerBriefs in Economics
ISBN 978-981-10-1096-5   ISBN 978-981-10-1097-2   (eBook)
DOI 10.1007/978-981-10-1097-2

Library of Congress Control Number: 2016938402

Printed on acid-free paper

This Springer imprint is published by Springer Nature
The registered company is Springer Science+Business Media Singapore Pte Ltd.

To Junko and Yuji.

Preface

The book presents an in-depth empirical analysis of consumer response to alternative policies for energy conservation. The main focus lies on innovative policy instruments that have drawn increasing attention from both academics and practitioners in energy conservation: critical peak pricing, conservation requests, in-home displays, and home energy reports. The book investigates the effects of these policy instruments on residential demand for electricity.

The data are drawn from a series of randomized field experiments undertaken by the Keihanna Eco-City Next-Generation Energy and Social Systems Demonstration Project Promotion Council for the fiscal years 2012–2013 in Japan, where serious concern about the shortage of electricity supply has emerged in the wake of the nuclear disaster in Fukushima. By applying econometric techniques to the quantitative analysis of residential consumption of electricity, the book demonstrates how consumers respond to the innovative instruments for energy conservation. It also offers new perspectives on using these policy instruments for energy conservation more effectively and indicates the potential for their practical implementation. I hope that this book is essential reading for energy specialists at both the academic and professional levels.

I am grateful to the financial support from JSPS KAKENHI Grant Number 15K03469, the CREST of Japan Science and Technology Agency, and the Musashi University Research Center. For the experimental data, I thank the Kansai Electric Power Company, the Mitsubishi Electric Corporation, and the Mitsubishi Heavy Industries, Ltd., which joined the Keihanna Eco-City Next-Generation Energy and Social Systems Demonstration Project Promotion Council.

I am indebted to valuable comments from Kenichi Akao, Toshi Arimura, Akira Hibiki, Hajime Katayama, Yoshifumi Konishi, Shunsuke Managi, Shigeru Matsumoto, Toru Namerikawa, Michael Roberts, Nori Tarui, and Kenko Uchida.

I also want to thank Kazuhiko Nishimura for his advice on the publication of this book. Furthermore, I thank the editorial team at Springer Japan who helped me with the publication of this book.

Last but not least, my deepest thanks go to my family. They always supported me and gave me the power to work on this book.

Tokyo   Isamu Matsukawa
March 2016

Contents

Chapter 1
Introduction

Abstract The main focus of this book lies on innovative policy instruments that have drawn increasing attention from both academics and practitioners in energy conservation: critical peak pricing, conservation requests, in-home displays, and home energy reports. The book investigates the effects of these policy instruments on residential demand for electricity, using the data drawn from a series of randomized field experiments for the fiscal years 2012–2013 in Japan, where serious concern about the shortage of electricity supply has emerged in the wake of the nuclear disaster in Fukushima. By applying econometric techniques to the quantitative analysis of residential consumption of electricity, the book demonstrates how consumers respond to the innovative instruments for energy conservation. It also offers new perspectives on using these policy instruments for energy conservation more effectively and indicates the potential for their practical implementation.

Keyword Energy conservation · Innovative policy instruments · Residential electricity consumption · Field experiments · Japanese households

This book investigates consumer behavior in response to innovative energy conservation policies. The worldwide need to promote energy conservation is growing and such conservation is expected to mitigate global warming by reducing greenhouse gas emissions. Conventional policy measures for energy conservation aim to promote the use of energy-efficient technologies in society. Examples include energy taxes, subsidies on energy-efficient investments, efficiency standards for energy-using durables, and energy auditing. The large body of literature on this topic has investigated the effectiveness of these policy measures.

The main focus of this book is on innovative policy instruments for energy conservation that have received increasing attention from both academics and practitioners: critical peak pricing (CPP), conservation requests (CRs), in-home displays (IHDs), and home energy reports (HERs). CPP raises energy prices during peak periods when the demand for energy is likely to almost reach the available energy supply capacity. CRs ask consumers to voluntarily reduce consumption during peak

© The Author(s) 2016
I. Matsukawa, *Consumer Energy Conservation Behavior After Fukushima*,
SpringerBriefs in Economics, DOI 10.1007/978-981-10-1097-2_1

periods without offering pecuniary incentives. IHDs and HERs provide consumers with detailed information on their energy usage along with energy conservation tips.

These innovative measures differ from conventional ones in that they promote the widespread use of user-friendly information technologies. Because of the increasing consumer awareness of the environment, the efficient use of energy-using durables such as electric appliances and cars is crucial to maximize consumer utility, subject to constraints on resources such as income, time, and effort. However, in sharp contrast to industrial customers whose well-informed staff efficiently monitor and control energy usage at factories and office buildings, consumers at home often find it difficult to determine the best energy use that maximizes their satisfaction subject to resource constraints.

The widespread use of information technologies has increasingly provided opportunities for these consumers to determine how to achieve the best use of energy without incurring large information search costs. For instance, consumers can make better choices by using the Internet to take into account a wide range of product attributes and prices. Homepages that use charts and animation to indicate the use of energy-using durables are examples of user-friendly interfaces that have been developed by energy utilities and appliance manufacturers attempting to facilitate consumer access to information on the efficient operation of their products and services.

The innovative energy conservation instruments take full advantage of information technologies available to consumers. Typically, CPP and CRs use e-mail and text messages to inform consumers of their electricity prices tomorrow and to ask consumers for voluntary reductions in their energy consumption. IHDs provide consumers with their real-time home usage of energy through information-feedback technologies that connect their residences to the control center. Consumers are easily able to view charts and animations of their HERs on their cellular phones and tablet displays.

This book investigates the effects of these innovative policy instruments on residential demand for electricity using Japanese data from a series of randomized field experiments after Fukushima. After the Great Earthquake caused a meltdown of the nuclear power plants in Fukushima in March 2011, the Japanese government tightened regulations on nuclear power plant safety. All nuclear power stations stopped operating until they proved that they met the new, much more stringent, safety standards. A serious concern over the supply shortage of electricity has emerged because nuclear power plants supplied approximately one-fourth of the electricity generated in Japan before Fukushima.

In response to the critical shortage of the supply of power, a series of field experiments were implemented to examine the effectiveness of innovative policies such as CPP, CRs, IHDs, and HERs. The focus of the experiments was on the residential sector that represented approximately one-fourth of electricity consumption in Japan. The Keihanna Eco-City Next-Generation Energy and Social Systems Demonstration Project Promotion Council, which includes the local government, utilities, and manufacturers of electric appliances and machinery, conducted randomized field experiments for the fiscal years 2012–2013. The subjects of the experiments were households living in Kyotanabe City, Kizugawa City, and Seika Town in the south of Kyoto, Japan.

The book demonstrates how consumers respond to CPP, CRs, IHDs, and HERs by applying econometric techniques to the quantitative analysis of the residential consumption of electricity. Chapter 2 provides an overview of energy conservation in Japan. Conventional policy measures, such as energy taxes, subsidies, labeling, and technological standards for energy efficiency, have contributed to the development and use of energy-efficient technologies. Indeed, energy efficiency has been consistently improved during the last three decades. This improvement of energy efficiency in Japan is remarkable among the developed countries. However, additional policy interventions are necessary to further constrain energy consumption because of the government's plan to reduce Japanese emissions of carbon dioxide (CO_2) and rising concern about electricity supply amid the lack of available nuclear power plants. Residential electricity saving is expected to mitigate the increasing concern about CO_2 emissions and electricity supply. Innovative instruments for energy conservation are expected to play an important role in further constraining the residential usage of electricity.

Chapter 3 measures the energy-conservation effects of CPP and CRs using the experimental data for the fiscal years 2012–2013. Experiments on CPP were conducted four times: summer 2012, winter 2012/2013, summer 2013, and winter 2013/2014. In these experiments, a limited number of weekdays were designated as "critical peak days" when regional demand for electricity is likely to almost reach the available capacity for the supply of electricity. Critical peak days were called on a day-ahead basis. The peak hours were from 1 p.m. to 4 p.m. in the summer and from 6 p.m. to 9 p.m. in the winter. A price of 65, 85, or 105 U.S. cents per kilowatt-hour (kWh) was applied to the electricity consumption of households assigned to the CPP treatment during the peak hours on critical peak days (1 U.S. dollar = 100 yen). Experiments on CRs were conducted twice: summer 2012 and winter 2012/2013. In these experiments, households assigned to the CR treatment were asked to reduce electricity consumption during the peak hours on critical peak days. This request was made to the households at approximately 8 p.m. on the day before each critical peak day through e-mail.

Both CPP and CRs were found to contribute to the reduction in households' daily electricity usage during peak hours. The empirical analysis of households' response to CPP and CRs builds on a "linear approximate almost ideal demand system" (LA/AIDS), which is consistent with utility maximization and is more general than such a traditional demand model as a constant elasticity of substitution model. The price elasticity of the peak electricity demand, which is defined as the ratio of a percent change in the demand to a percent change in the price, ranged from 0.157 to 0.389 in absolute terms at the sample average. The estimates of the price elasticity depended on electricity prices and seasons. These estimates exceed those found in the previous studies on CPP. CRs reduced the electricity demand for peak hours by 4.0 % during summer 2012 and by 5.1 % in winter 2012/2013. These estimates of the peak-reducing effects of CRs are lower than those reported in the United States.

Chapter 4 investigates how acquiring information from an IHD affects electricity consumption through attention and learning using data on the frequency at which each household uses an IHD in summer 2012 and winter 2012/2013. Households in the treatment group are able to view a graph of their electricity consumption in half-hour increments in real time on IHDs at any time during the experiment. Providing an IHD is a promising policy intervention that corrects for the consumption biases associated with inattention and limited information-processing capacity by heightening attention and learning. The immediate effect of providing an IHD is heightened household attention to information on consumption, and the repetition of attention is expected to improve households' capacity to process information in the long run.

The econometric analysis of the panel data on daily electricity consumption indicates statistically significant and persistent effects of IHD use on residential electricity consumption. The increase in IHDs' effects and households' experience using IHDs imply that households' capacity to process information could be improved by the repetition of attention to electricity information. The empirical evidence offered by Chap. 4 also indicates that providing an IHD increased household electricity consumption. Thus, providing households with IHDs, which was found in previous studies to be an effective policy instrument for energy conservation, could have an adverse effect on energy conservation. However, an interactive effect of IHD provision and CPP implies that providing an IHD together with pecuniary incentive schemes could be effective in energy conservation.

Chapter 5 measures the energy-saving effects of HERs using data from the experiment in 2013. HERs provide households with personalized tips for energy conservation and compare households' energy usage with that of similar neighbors. Specifically, the professional staff visited each household during May 9 through July 5, 2013, in order to explain these informational items with a leaflet. The leaflet compared the hourly electricity usage of each household with that of the similar neighbors. The neighbor comparisons categorized households as "energy-using," "average," or "energy-saving" in order to provide social norm information. These comparisons are expected to induce households to conserve energy.

The energy-saving effects of HERs, which are indicated by a panel-data analysis of half-hourly electricity consumption on weekdays during May 9 through July 5, 2013, depend on whether the electricity contracts that households enter into with the local utility are standard or all-electric contracts. For standard-contract households that used gas or kerosene for space heating and water heating, HERs did not contribute to a reduction in electricity consumption. In contrast, HERs reduced the electricity consumption of households with all-electric contracts. The electricity-saving effect of HERs on all-electric households, which ranged from 4.0–8.7 %, was found in each category of electricity usage. Overall, there was no indication of a so-called "boomerang" problem, which raises electricity usage of households categorized as "energy-saving." The electricity-saving effect of HERs was found to become larger during the morning and evening of weekdays. This effect could be persistent over several weeks after providing HERs for some groups of households.

Chapter 6 develops a forecast to assess the regional impact of CPP, IHDs, and HERs using the empirical evidence of consumer response to energy conservation policies. The forecast offers perspectives on more effective use of these innovative instruments for energy conservation and indicates the potential for their practical implementation. In the forecast of CPP together with IHDs, for the peak hours from 1 p.m. to 4 p.m. in summer 2013 and from 6 p.m. to 9 p.m. in winter 2013/2014 on critical peak days, a price of 65, 85, or 105 cents per kWh is assumed to have been applied to the electricity consumption of 10,150,000 households contracting standard and all-electric in the Kansai region where the experimental site is located. These households are assumed to have used IHDs once each critical peak day. In the forecast of HERs, households contracting all-electric in the Kansai region are assumed to have received reports and their detailed explanation from the professional staff at the beginning of June 2013. The report is assumed to have consisted of personalized tips for energy conservation and neighbor comparisons of electricity usage. All-electric households in the region are assumed to have been categorized as "energy-saving," "average," or "energy-using" based on their historical usage of electricity.

The combination of CPP and IHDs would have reduced residential electricity usage during the peak period in the Kansai region by approximately 15.0 % in summer 2013 and 19.7 % in winter 2013/2014. Overall, these effects of CPP together with IHDs on electricity savings would be far larger than those of the government's call for voluntary reduction and would contribute to the reduction in the costs of regional electricity supply during the peak period. The impact of CPP on the region extends to the possible prevention of outages during peak hours on critical peak days and to CO_2 emission reductions associated with the operation of oil-fired generation plants. HERs are expected to reduce residential electricity consumption, thus, by applying HERs to all-electric households in the Kansai region in June 2013, a household would have saved 4.9–8.3 %. The cost effectiveness of HERs, which is defined as the implementation costs of HERs per electricity saving, would range from 0.77 to 1.30 dollars per kWh if the implementation costs include both leaflet and staff costs.

Chapter 2
Energy Conservation in Japan

Abstract A serious concern about the supply shortage of electricity after the nuclear disaster in Fukushima, together with a growing concern about global warming, has enhanced the need to promote energy conservation in Japan. The residential sector, which is the focus of this book, has increased its share of total energy consumption for the past two decades in Japan. Electricity represents approximately half of the residential energy consumption, and is a key factor in promoting energy conservation in the residential sector. Conventional policy measures such as energy taxes, subsidies, labeling, and technological standards for energy efficiency have contributed to the development and use of energy-efficient technologies, but additional policy interventions are necessary to further constrain electricity consumption. Innovative energy conservation instruments such as critical peak pricing, conservation requests, in-home displays, and home energy reports are expected to mitigate the increasing concern about electricity supply and CO_2 emissions by constraining the residential usage of electricity in the future.

Keywords Energy taxation · Subsidization · Labeling · Energy efficiency · Standards · CO_2 emissions

2.1 Energy Conservation Policies in Japan

The Japanese government has enforced the Energy Conservation Law (ECL), which was enacted in 1979 in the wake of the second oil crisis, to implement regulatory measures regarding energy conservation. The ECL obliges firms that consume large amounts of energy to disclose annual energy consumption and to submit long-term plans on conservation (JANRE 2011). These firms are found primarily in the manufacturing and transportation sectors. The ECL also requires manufacturers and importers of energy-using durable goods such as passenger cars, air-conditioners, and television sets, to meet so-called "Top Runner Standards," which are energy efficiency standards that are based on the best available technologies. To heighten consumers' awareness of energy efficient

© The Author(s) 2016   7
I. Matsukawa, *Consumer Energy Conservation Behavior After Fukushima*,
SpringerBriefs in Economics, DOI 10.1007/978-981-10-1097-2_2

Table 2.1 Energy taxation by the Japanese government

Taxation	Fuel	Tax rate
Gasoline tax	Volatile oil (gasoline)	48.6 cents/L
Oil and gas tax	LPG for the car	17.5 cents/kg
Aviation fuel tax	Aviation fuel	18.0 cents/L
Petroleum and coal tax (inclusive of tax for climate change mitigation)	Crude oil and petroleum products	28.00 dollars/kL
	LPG and LNG	18.60 dollars/ton
	Coal	13.70 dollars/ton
Electric power development promotion tax	Electricity	3.75 dollars per 1,000 kWh

Note Tax rates are converted to U.S. currency (1 U.S. dollar = 100 yen)
Source Japan Ministry of the Environment. (http://www.env.go.jp/en/policy/tax/env-tax/20120814a_ertj.pdf, accessed January 2016)

products, the government encourages manufactures of energy-using durables to label energy consumption and energy efficiency on their products. Energy efficiency standards are also applied to commercial buildings and residential dwellings.

Together with a command-and-control approach, the government has adopted economic instruments for energy conservation (Arimura and Iwata 2015). Taxation and subsidization are the main economic instruments used in this regard. Table 2.1 summarizes energy taxation associated with the Japanese government. Tax rates are converted to U.S. currency (1 U.S. dollar = 100 yen). Gasoline tax, oil and gas tax, and aviation fuel tax are imposed on energy usage for transportation. The petroleum and coal tax is applied to the consumption of crude oil, petroleum products, gas (liquefied natural gas, LNG, and liquefied petroleum gas, LPG), and coal. This tax includes a charge for climate change mitigation. Finally, consumers pay the electric power development promotion tax for their electricity usage.

The government has also provided firms with subsidies for energy conservation measures. An example of this is the energy-efficient replacement of facilities at manufacturing plants. The subsidy provides for up to one-third of the costs for each facility replacement. Firms installing new energy-efficient facilities or purchasing equipment that meets efficiency standards can also receive a subsidy for interest payments. Additionally, subsidization is also available for the construction of highly efficient buildings and dwellings.

The important policy question is whether these conservation measures have contributed to an improvement in energy efficiency in Japan. Figure 2.1 illustrates energy intensity, which is defined as the ratio of total primary energy requirements (measured in ton of oil equivalent [toe]) to the gross domestic product (GDP) (measured in 100 million yen, 2005 price), for the period 1975–2012. Energy intensity indicates how efficiently energy is used in the economy. During the period of 1975–2012, energy intensity consistently decreased, dropping by approximately 38.9 % overall. This improvement of energy efficiency in Japan is

Fig. 2.1 Energy intensity in Japan, 1975–2012. *Note* Energy intensity is the ratio of total primary energy requirements (measured in ton of oil equivalent) to GDP (measured in 100 million yen, 2005 price). *Source* EDMC (2015)

remarkable among the developed countries. Table 2.2 compares energy consumption among selected OECD countries in 2012. Among six countries listed in the table, Japan is the second largest consumer of energy, and its share of electricity is the largest. The international comparison of energy intensity indicates that Japan is less energy-intensive than the OECD average. It uses energy more efficiently than France, Germany, and the United States. These findings indicate that energy conservation policies, amongst other factors, are key contributors to the improvement of energy efficiency in Japan.

Further conservation of energy is necessary for Japan, however, since its carbon dioxide (CO_2) intensity, which is measured by the ratio of CO_2 emissions to GDP,

Table 2.2 Energy consumption by selected OECD countries, 2012

	Japan	Italy	France	Germany	United Kingdom	United States	OECD total
Energy consumption (Mtoe)	452	159	252	313	192	2,141	5,250
—Electricity share (%)	25.7	20.8	24.1	20.5	21.4	22.4	22.3
Energy intensity (toe/million US$, 2010 price)	82	79	96	91	83	137	116
CO_2 intensity (CO_2 ton/million US$, 2010 price)	220	176	124	211	201	328	267

Notes Mtoe stands for million ton of oil equivalent. Energy intensity is the ratio of total primary energy requirements (measured in ton of oil equivalent [toe]) to GDP (measured in million U.S. dollars, 2010 price). CO_2 intensity is the ratio of total CO_2 emissions to GDP
Source EDMC (2015)

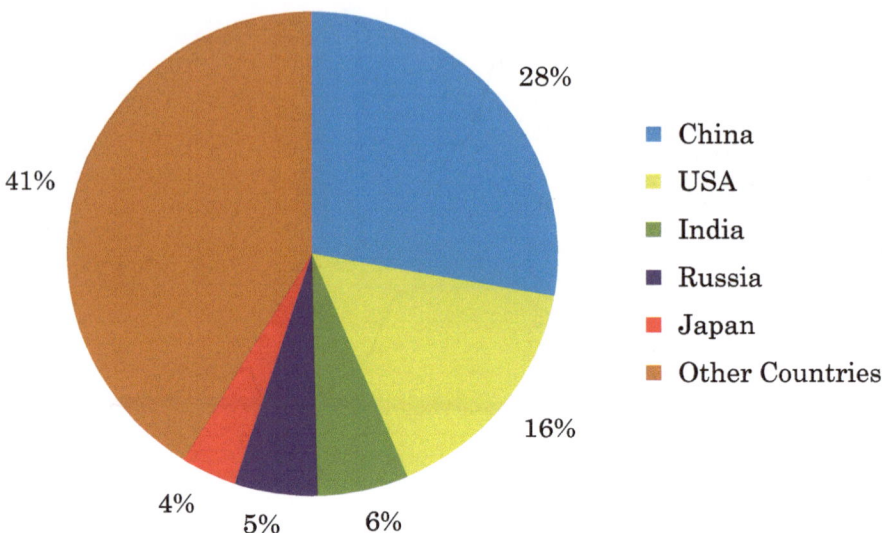

Fig. 2.2 Shares of CO_2 emissions by country, 2012. *Source* EDMC (2015)

is larger than that in Italy, France, Germany, and the United Kingdom, as shown in Table 2.2. In fact, as Fig. 2.2 shows, Japan was the fifth largest emitter of CO_2 in the world as of 2012, following China, the United States, India, and Russia. Further conservation of energy would contribute to constraining CO_2 emissions, because more than 90 % of CO_2 emissions originate from energy consumption in Japan (EDMC 2015).

The Japanese government has attempted to constrain CO_2 emissions, which represent more than 90 % of total emissions of greenhouse gases in the country. Despite its efforts, CO_2 emissions have steadily increased in Japan, with an annual average growth rate of approximately 0.6 % during the period of 1990–2013 (EDMC 2015). By using the carbon sink in domestic forests and the Kyoto Mechanisms, the government has achieved its targeted reduction of greenhouse gases from the Kyoto Protocol: a 6 % reduction in annual average emissions below the 1990 level during a five-year commitment period (2008–2012). Recently, the government submitted a plan to the United Nations prior to the Paris climate conference in 2015 to reduce greenhouse gas emissions by 26 % from 2013 levels by the year 2030.

2.2 Energy Conservation in the Residential Sector

The residential sector, which is the focus of this book, has increased its share of total energy consumption for the past two decades. Along with this increase in energy consumption, the residential sector raised its share of CO_2 emissions from

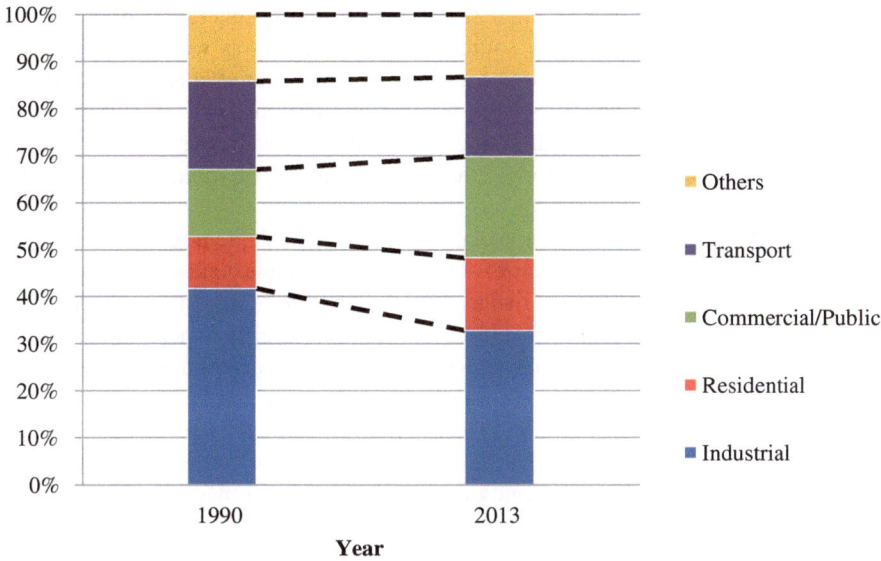

Fig. 2.3 Shares of CO_2 emissions by sector in Japan. *Source* EDMC (2015)

11 % in 1990 to 16 % in 2013, as shown by Fig. 2.3. This is in sharp contrast with the industrial sector that dropped its share of CO_2 emissions from 42 % in 1990 to 33 % in 2013. Constraining CO_2 emissions from residential energy usage is crucial for the achievement of the plan submitted to the United Nations to reduce CO_2 emissions in Japan.

Table 2.3 provides a breakdown of 2013 residential energy use in Japan. The end-use category "other uses" includes lighting equipment, refrigerators, television sets, clothes washers, dryers, dishwashers, electronic equipment, and numerous small electric appliances. Except for these "other uses" whose energy source is electricity, water heating is the largest among the major end-use categories. Electricity represents approximately half of the residential energy consumption.

Table 2.3 Residential energy consumption by fuel and end-use categories, 2013 (1,000 kilocalories per household)

	Space heating	Air conditioning	Water heating	Cooking	Other uses	Total (%)
Electricity	377	223	347	206	3,281	4,434 (49.1)
Gas	494	0	1,688	566	0	2,748 (30.5)
Kerosene	1,432	0	405	0	0	1,837 (20.4)
Total (%)	2,303 (25.5)	223 (2.5)	2,440 (27.1)	772 (8.6)	3,281 (36.3)	9,019 (100.0)

Source EDMC (2015)

Thus, saving electricity is the main target for residential energy conservation in Japan. Gas is the primary energy source for water heating as well as for cooking. Space heating occupies roughly one-fourth of total residential energy demand, and kerosene is the primary energy source for this end-use category. Electricity is the only energy source for air conditioning, which has a small consumption share of 2.5 % among all end-use categories, though the number of room air conditioners was 2.8 per household in 2013.

Residential energy conservation depends on technical, demographic, climatic, economic, and psychological factors. These factors include end-use technologies, dwelling and household characteristics, climate conditions, income, energy prices, and environmental consciousness. Given these factors, energy conservation policies of the Japanese government have focused on the development and use of energy-efficient technologies by regulatory and economic measures.

Table 2.4 summarizes the main conservation measures for the residential sector in Japan. Gasoline-fueled cars, energy appliances, and dwellings are subject to regulation on energy efficiency standards. These standards reflect the best available technologies. The fuel economy of gasoline-fueled cars under the current standard is higher than that in 1995 by 22.8 %. Energy efficiency improvements under the current standards substantially differ across energy appliances: from the 1.9 % improvement relative to the 2000 level in the thermal efficiency of gas heaters, to the 67.8 % improvement relative to the 1997 level in the coefficient of performance of air conditioners. Because of standards on heat loss coefficients and thermal insulation materials, annual energy consumption of the typical house that is subject to the current standard is less than the annual energy consumption of non-insulated houses by 60.7 %.

Labeling enables consumers to be aware of the energy efficiency of each electrical appliance. Labeling on such appliances as air conditioners, television sets, refrigerators, and fluorescent lights indicates annual electricity use relative to the efficiency standards. Subsidies for energy-efficient appliances take a form of

Table 2.4 Energy conservation measures for the residential sector in Japan

Measures	Items	Details
Efficiency standards	Gasoline-fueled cars	Fuel economy of 15.1 km per liter (22.8 % improvements in energy efficiency)
	Energy appliances	Energy efficiency improvement ranging from 1.9 to 67.8 %
	Dwellings	Heat loss coefficients and thermal insulation materials
Labeling	Electrical appliances	Annual electricity use relative to the efficiency standards
Tax deductions and subsidies	Electrical appliances	Subsidies for energy-efficient appliances
	Dwellings	Tax deductions, low-interest mortgages, and subsidies for newly built or renovated energy-efficient houses

Source JANRE (2011)

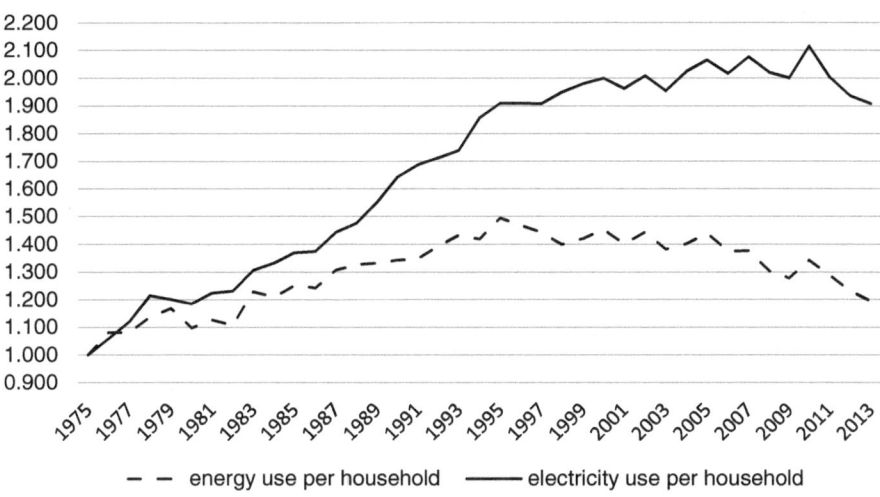

Fig. 2.4 Energy use per household in Japan, 1975–2013 (1975 = 1). *Source* EDMC (2015)

electronic points that can be used to purchase designated goods or can be exchanged for merchandise vouchers. Tax deductions, low-interest mortgages, and subsidies are also available to consumers who invest in newly built or renovated energy-efficient houses.

These conservation measures have contributed to constraining energy use in the residential sector for the last two decades. Indeed, since the late 1990s, energy consumption per Japanese household has gradually declined, as shown by Fig. 2.4. This decrease in energy use per household reflects the impact of the energy conservation measures implemented by the Japanese government. Before the nuclear disaster in Fukushima that occurred in 2011, however, electricity use per household continued to increase. Although electricity consumption per household declined in the wake of the nuclear disaster in Fukushima, it is not clear whether this downturn trend will be temporary or persistent. An additional policy intervention is necessary to further constrain electricity use in the residential sector because of a rising concern about the security of the electricity supply in Japan.

2.3   Security of the Electricity Supply After Fukushima

Along with increased CO_2 emissions, a rising concern about the secure supply of electricity has raised residential energy conservation to an urgent need in the wake of the nuclear disaster that occurred in March 2011 in Fukushima. The concern about a secure supply of electricity after Fukushima emerged as the demand for electricity came precariously close to the available electricity supply capacity in several regions.

Table 2.5 Regional electricity supply and demand in Japan on the day of the highest daily demand in the summer of 2013

Utility	Supply (million kilowatts)	Demand (million kilowatts)	Reserve margin (%)	Maximum temperature (°C)
Hokkaido	5.4	4.5	21.1	31.0
Tohoku	15.0	13.2	13.6	32.6
Tokyo	54.9	50.9	7.9	35.1
Chubu	27.3	26.2	4.0	38.4
Kansai	29.4	28.2	4.3	37.0
Hokuriku	5.5	5.3	5.1	36.3
Chugoku	11.7	11.1	5.0	35.4
Shikoku	5.8	5.5	5.0	35.5
Kyushu	17.0	16.3	4.3	36.5
Okinawa	2.1	1.5	36.2	33.6

Note The reserve margin is the ratio of excess supply to the maximum demand for electricity
Source JMETI (2013)

Table 2.5 presents regional electricity supply and demand on the day when the demand was the highest in the summer of 2013. On that day, the maximum ambient temperature ranged from 31.0–38.4 °C, and it exceeded 35.0 °C in seven regional utilities out of ten. As of 2013, ten vertically integrated, privately owned electric utilities were to supply electricity to their regions. As shown by Table 2.5, in six regional utilities, the reserve margin, the ratio of excess supply to the maximum demand for electricity, was much lower than 8 %, which is considered the minimum level to ensure a secure regional supply of electricity in Japan.

The insecure supply of electricity was primarily due to the ceased operation of nuclear power plants after the nuclear disaster in Fukushima. Figure 2.5 compares the electricity supply share of each fuel type in Japan between two periods: before and after the nuclear disaster. Fossil fuels include oil, gas (LNG), and coal. Renewables include wind, solar, and geothermal power. Figure 2.5 indicates that nuclear power plants, which generated approximately one-fourth of total electricity supply in Japan before the disaster, represented only 1.0 % of electricity supply in 2013. Power plants using fossil fuels substituted for nuclear power plants after the disaster. In fact, almost all nuclear power plants ceased operations in 2013, and no nuclear power plants remained operational by 2014. Safety regulations of nuclear power plants require nuclear power stations to stop operating until they have met new safety standards, which are more stringent than before the disaster.

To secure electricity supply, the Japanese government imposed mandatory restrictions of electricity use on industrial and commercial customers, contracting 500 kW or more in the Tohoku and Kanto regions (JCS 2011). The Tohoku region includes Fukushima where the nuclear disaster occurred. The Kanto region, which includes the Tokyo metropolitan area, is adjacent to the Tohoku region. The industrial and commercial customers who contracted 500 kW or more in these regions were obliged to reduce electricity use for peak hours in the summer of 2011

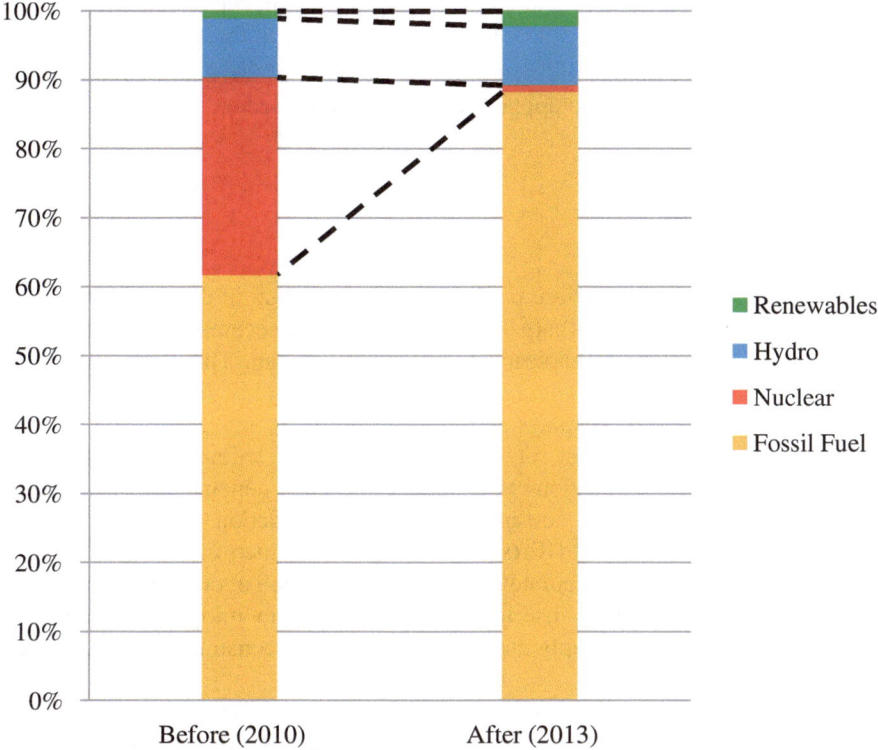

Fig. 2.5 Kilowatt-hour shares of electricity generation by fuel in Japan, before and after the nuclear disaster in Fukushima. *Notes* Fossil fuels include oil, gas (LNG), and coal. Renewables include wind, solar, and geothermal power. *Source* Federation of Electric Power Companies of Japan (http://www.fepc.or.jp/, accessed January 2016)

by 15 % from usage levels in the summer of 2010. The government also called for a voluntary reduction in the electricity use of other customers in the Tohoku and Kanto regions. Voluntary electricity saving was also called for in the Kansai region, which includes one of the three metropolitan areas in Japan.

These mandatory and voluntary restrictions on electricity use achieved a substantial reduction in electricity demand during peak hours in the summer of 2011 (JCS 2011). The maximum demand for electricity was reduced by 18 % relative to the previous year in the Tohoku region. The reductions in maximum demand for electricity were 19 and 8 % in the Kanto and Kansai regions, respectively. These reductions in the peak electricity demand together with the increased trade of electricity across regions prevented electricity supply from being interrupted.

In November 2011, the Japanese government decided that additional measures were required in its "action plan" for the secure supply of electricity (JCS 2011). These measures for the residential sector aimed to promote home energy management systems (HEMS), smart meters, storage batteries, energy-efficient appliances and dwellings, photovoltaics, and fuel cells. The government also suggests that electricity pricing, which contributed to the reduction in the peak electricity demand of industrial sectors in the summer of 2012 and 2013 (Isogawa and Ohashi 2015), should be applied to the residential sector.

2.4 Conclusion

Residential energy conservation, particularly the reduction in the peak demand for electricity, is expected to mitigate the concern about the secure supply of electricity amid the lack of available nuclear power plants in Japan. The innovative instruments for energy conservation (i.e., critical peak pricing [CPP], conservation requests [CRs], in-home displays [IHDs], and home energy reports [HERs]) are expected to play an important role in the government's action plan for the secure supply of electricity. CPP attempts to reduce the peak demand for electricity by pricing, while CRs rely on consumers' voluntary reduction of electricity usage during peak hours. IHDs and HERs could complement smart meters and HEMS by providing consumers with accurate and timely information about electricity use and energy conservation. These instruments are expected to mitigate the increasing concern about electricity supply and CO_2 emissions by constraining the electricity use of households.

Focusing on conventional instruments such as time-of-use pricing, the previous studies (Matsukawa et al. 2000; Matsukawa 2001, 2004, 2011) provide evidence that the use of innovative instruments could be potentially effective for Japanese consumers. The residential sector was found to modestly adjust demand for electricity in response to a temporal variation in electricity prices, which could suggest the effectiveness of CPP (Matsukawa et al. 2000; Matsukawa 2001). Additionally, the residential sector was found to slightly reduce electricity consumption in response to the provision of information about electricity usage at home and energy-conservation tips (Matsukawa 2004, 2011), which could suggest the effectiveness of IHDs and HERs.

The following chapters in this book focus on the effects of innovative instruments for residential energy conservation that have not been investigated by the previous studies on Japanese consumers. These effects are investigated during a series of randomized field experiments. The experiments were conducted as part of the Keihanna Eco-City Next-Generation Energy and Social Systems Demonstration Project, which was subsidized by the Japanese government and aimed to evaluate alternative systems for energy management concentrating on energy conservation, reduction in the peak demand for energy, and balancing energy supply and demand (NEPC 2016). These energy management systems targeted residential communities and commercial buildings where HEMS, electric vehicles, photovoltaics, and storage batteries were installed. In addition to these systems, the project also conducted a series of experiments to determine how CPP, CRs, IHDs, and HERs affected the electricity consumption of households.

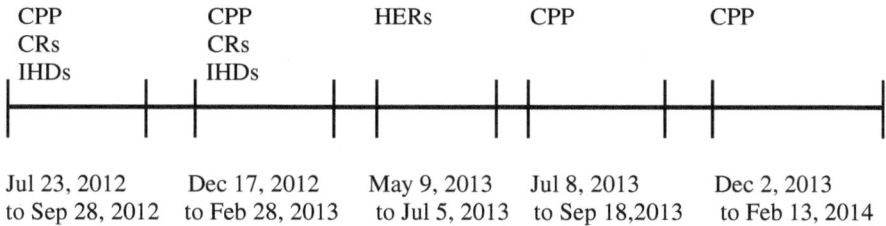

Fig. 2.6 Timeline of the field experiments on which Chaps. 3–5 focus

Using the data from the field experiments, Chaps. 3–5 examine the effects of CPP, CRs, IHDs, and HERs on residential usage of electricity. Figure 2.6 illustrates the timeline of the field experiments on which these chapters focus. Chapter 3 focuses on CPP and CRs whose experiments were conducted in summer 2012 (from July 23 to September 28) and in winter 2012/2013 (from December 17 to February 28). It also focuses on CPP experiments in summer 2013 (from July 8 to September 18) and in winter 2013/2014 (from December 2 to February 13). Chapter 4 focuses on IHDs from which participating households could obtain electricity information at any time during the first and second experiments on CPP and CRs. Chapter 5 focuses on HERs participating households received during May 9 through July 5 in 2013.

References

Arimura T, Iwata K (2015) An evaluation of Japanese environmental regulations, Springer

EDMC [Energy Data and Modelling Center, Institute of Energy Economics, Japan] (2015) Handbook of Japan's and world energy and economic statistics, Energy Conservation Center, Japan

Isogawa D, Ohashi H (2015) Effects of load adjustment contracts on industrial electricity use in Japan. Research Institute of Economy, Trade and Industry, Discussion Paper 15-J-053 (in Japanese)

JANRE [Japan Agency of Natural Resources and Energy] (2011) Energy conservation policies of Japan. November 7

JCS [Japan Cabinet Secretariat] (2011) Energy supply–demand stabilization action plans, Energy and Environment Council, November (in Japanese). http://www.cas.go.jp. Accessed January 2016

JMETI [Japan Ministry of Economy, Trade and Industry] (2013) Report of the electricity supply–demand verification subcommittee, October (in Japanese)

Matsukawa I (2001) Household response to optional peak-load pricing of electricity. J Regul Econ 20(3):249–267

Matsukawa I (2004) The effects of information on residential demand for electricity. Energy J 25(1):1–17

Matsukawa I (2011) How does information provision affect residential energy conservation? Evidence from a field experiment. Energy Stud Rev 18(1):1–19

Matsukawa I, Asano H, Kakimoto H (2000) Household response to incentive payments for load shifting: a Japanese time-of-day electricity pricing experiment. Energy J 21(1):73–86

NEPC [New Energy Promotion Council] (2016) Japan smart city portal, http://jscp.nepc.or.jp/en/. Accessed January 2016

Chapter 3
Consumer Response to Critical Peak Pricing of Electricity and Conservation Requests

Abstract Using data from a series of field experiments for the fiscal years 2012–13, this chapter investigates consumers' electricity use in response to critical peak pricing (CPP) and conservation requests (CRs). CPP stimulates consumers' extrinsic motivation to conserve energy by applying higher prices to electricity usage during peak periods. This is a plausible approach since electricity pricing affects the monetary reward of conservation and higher prices create conservation that is more beneficial. By asking households to voluntarily reduce consumption during peak periods, CRs stimulate intrinsic motivations, such as altruism and the "warm glow," which enhance consumers' utility through energy conservation. The empirical results based on a linear approximate almost ideal demand system imply that both CPP and CRs contributed to reducing electricity usage during peak periods. In absolute terms, the elasticity of the peak electricity demand with respect to the electricity price at the sample average ranged from 0.157 to 0.389. These estimates of price elasticity exceeded those found in previous studies on CPP. CRs reduced electricity demand during peak hours by 4.0–5.1 %. These peak-reducing estimates of CRs were lower than the effects of public appeals to conserve energy in the United States.

Keywords Critical peak pricing · Conservation requests · Price elasticity · Almost ideal demand system · Panel data

3.1 Introduction

This chapter investigates consumer use of electricity in response to two policy instruments for energy conservation: critical peak pricing (CPP) and conservation requests (CRs). CPP applies higher prices to consumption during peak periods; its effects have been investigated in recent years by several field experiments (Wolak 2011; Faruqui et al. 2014; Fenrick et al. 2014; Jessoe and Rapson 2014). CRs ask households to voluntarily reduce consumption during peak periods without offering pecuniary incentives (Reiss and White 2008; Holladay et al. 2015).

© The Author(s) 2016

I. Matsukawa, *Consumer Energy Conservation Behavior After Fukushima*, SpringerBriefs in Economics, DOI 10.1007/978-981-10-1097-2_3

Although both policy interventions are expected to reduce peak period electricity usage, their effects on electricity consumption depend on two different types of motivation: extrinsic and intrinsic (Benabou and Tirole 2006). CPP stimulates households' extrinsic motivation to conserve energy because electricity pricing affects the monetary reward of conservation, and higher prices make conservation more beneficial. CRs stimulate intrinsic motivations such as altruism and the "warm glow," which enhance households' utility through energy conservation.

Using data from a series of field experiments, this chapter aims to measure the effects of CPP and CRs on the residential consumption of electricity during peak hours. The Keihanna Eco-City Next-Generation Energy and Social Systems Demonstration Project Promotion Council (hereafter, referred to as the Keihanna Eco-City Promotion Council) conducted four experiments during the fiscal years 2012–13 in southern Kyoto, Japan. These experiments correspond to "framed field experiments," which typically use experimental participants from the market of interest and incorporate important elements within the context of the naturally occurring environment with respect to the commodity, task, stakes, and information set of the subjects (List and Price 2013). The experimental site was located in one of the four regions where "smart city" projects were conducted (NEPC 2016). These projects received subsidies from the Ministry of Economy, Trade and Industry. They selected CPP and CRs as promising policy instruments for demand response, aiming to constrain electricity usage during peak hours by stimulating consumers' motivation to save energy.

Ito et al. (2015) measured how CPP and CRs (or moral suasion in their terms) affected residential consumption of electricity using half-hourly data from the experiment of the Keihanna Eco-City Promotion Council during the summer of 2012 to the spring of 2013. They found that CPP reduced the peak demand for electricity during the experiment. They also found that the impact of CPP was persistent while CR effects diminished quickly as the experiment proceeded.

This chapter differs from the previous literature on CPP in an important way. It applies a linear approximate almost ideal demand system (LA/AIDS) model to the analysis of residential time-of-use electricity consumption. An LA/AIDS model could well approximate consumers' utility function, thereby enabling a welfare analysis to be consistent with standard economic theory of utility maximization. This utility-consistent feature of an LA/AIDS model stands in contrast to a reduced-form model of electricity demand, which has been used in the literature on CPP (Wolak 2011; Jessoe and Rapson 2014; Ito et al. 2015). Moreover, in comparison to a constant elasticity of substitution (CES) model, which is assumed in Faruqui et al. (2014) and Fenrick et al. (2014), an LA/AIDS model is more general in the sense that elasticities of substitution between peak and off-peak hours differ across consumers.

The rest of this chapter proceeds as follows. Section 3.2 describes the design of the experiments and data obtained from the experiment. Section 3.3 presents an empirical model of residential demand for time-of-day electricity consumption. This section also discusses the estimation results of the model, and the policy implications of the empirical findings. Section 3.4 concludes this chapter. The appendix section describes the estimation procedure for measuring the price elasticity of total electricity demand.

3.2 Experimental Design and Data

During the fiscal years 2012–13, the Keihanna Eco-City Promotion Council conducted four experiments in which participating households in a southern area of Kyoto, Japan, were assigned to either control or treatment groups (i.e., CR, CPP, and CPP+ home energy report [HER] groups). The experimental site was located in the Kansai Science City. The Kansai Science City has been built in the areas belonging to the Kyoto, Osaka, and Nara prefectures (Fig. 3.1). As shown by Table 3.1, in three treatment groups, households received either CPP or CRs in the first and second experiments. They received CPP in the third and last experiments. Before these experiments, the CR and CPP+ HER groups received HERs. The details of HERs will be discussed in Chap. 5.

The first experiment was implemented in the summer, from July 23, 2012, to September 28, 2012 (68 days). Following that, a second was conducted in the winter, from December 17, 2012, to February 28, 2013 (74 days). The third experiment was done in the summer, from July 8, 2013, to September 18, 2013 (73 days), and the last one was in the winter, from December 2, 2013, to February 13, 2014 (74 days). All experiments were based on the same households. The area's climate during the first and third experiments was hot and humid; the maximum ambient temperature often exceeded 30 °C in summer (see Fig. 3.2). As shown in Fig. 3.2, the winter was mild. The minimum ambient temperature averaged just below 0 °C during the second and fourth experiments.

Fig. 3.1 Location of the Kansai Science City. *Source* Japan Ministry of Land, Infrastructure, Transport and Tourism (http://www.mlit.go.jp/crd/daisei/daikan/gaiyo_e.html, accessed January 2016)

Table 3.1 Experimental design

	Experiment	Control group	CR group	CPP group	CPP + HER group
CPP	First and second	No	No	Yes	Yes
	Third and last	No	No	Yes	Yes
CRs	First and second	No	Yes	No	No
	Third and last	No	No	No	No

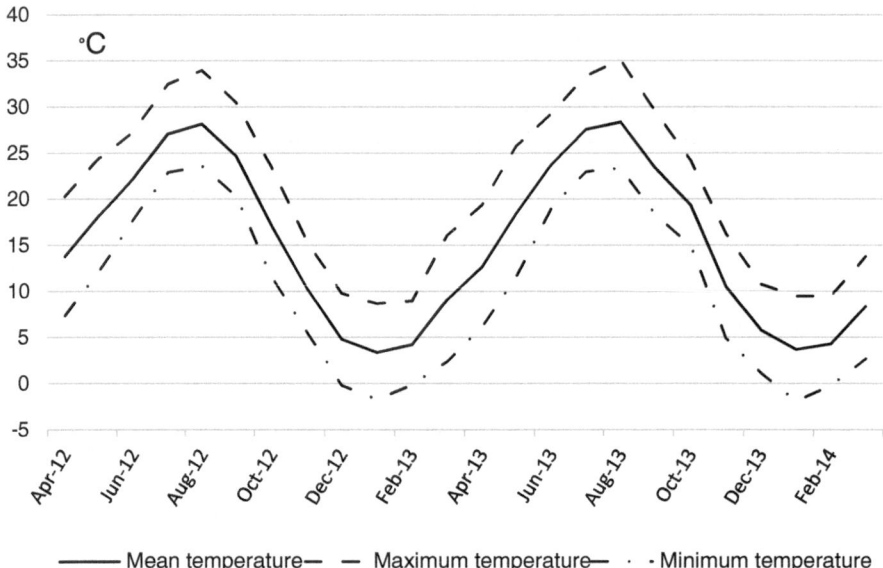

Mean temperature— — Maximum temperature— · · Minimum temperature

Fig. 3.2 Ambient temperature during experimentation. *Source* Japan Meteorological Agency (http://www.jma.go.jp, accessed January 2016)

3.2.1 CPP Experiments

Households allocated either to the CPP or CPP + HER groups received higher electricity prices during peak hours on days when the regional demand for electricity was likely to almost reach the available capacity for electricity supply. These days are often referred to as "critical peak days." In the experiment, a limited number of weekdays were designated as critical peak days and were called on a day-ahead basis. The peak hours were from 1 p.m. to 4 p.m. in the summer, and from 6 p.m. to 9 p.m. in the winter.

A price of 65, 85, or 105 U.S. cents per kilowatt-hour (kWh) was applied to electricity consumption during the peak hours on critical peak days (1 U.S. dollar = 100 yen). On weekdays not designated as critical peak days, a price of 45 cents/kWh was applied to electricity consumption during the peak hours. On weekdays, a price of 25 cents/kWh was applied to electricity consumption during

Table 3.2 Number of critical peak days by ambient temperature

	Temperature(°C)	65 cents/kWh	85 cents/kWh	105 cents/kWh	Total
Summer in 2012, 15 days	Below 33.0	2	2	1	5
	33.0–34.9	2	3	3	8
	35.0 or above	1	0	1	2
Winter in 2012/13, 24 days	5.0 or above	1	2	2	5
	2.0–4.9	3	2	2	7
	Below 2.0	4	4	4	12
Summer in 2013, 16 days	Below 33.0	1	1	1	3
	33.0–34.9	2	3	2	7
	35.0 or above	3	1	2	6
Winter in 2013/14, 21 days	5.0 or above	2	2	2	6
	2.0–4.9	2	1	1	4
	Below 2.0	3	4	4	11

Note Ambient temperature in summer (winter) is the maximum (minimum) temperature during the peak period, which was from 1 p.m. to 4 p.m. in summer and from 6 p.m. to 9 p.m. in winter
Source The Keihanna Eco-City Next-Generation Energy and Social Systems Demonstration Project Promotion Council

the off-peak hours. This price was also applied to electricity consumption during both peak and off-peak hours on weekends and holidays. Households allocated to either the control or the CR group received 25 cents/kWh at any time during the experiment. The experiment's maximum electricity price of 105 cents/kWh is a little more than four times the off-peak price of 25 cents/kWh. This price difference is smaller than that in the CPP literature (Wolak 2011; Faruqui et al. 2014; Fenrick et al. 2014; Jessoe and Rapson 2014).

In the first two experiments, there were five critical peak days for each price in the summer and eight for each price in the winter. Six critical peak days for 65 cents/kWh and five of these days for 85 or 105 cents/kWh were applied to the treatment in the third experiment (summer). There were seven critical peak days for each price in the last experiment (winter). As shown in Table 3.2, the price applied did not depend on ambient temperature. The electricity price of 45 cents/kWh was applied to peak usage on weekdays except for critical peak days. Households in the treatment were informed of the maximum number of critical peak days before each experiment began. These households were informed of the electricity price at approximately 8 p.m. the day before each critical peak day by e-mail, which they accessed on their cellular phones and personal computers. The notice of the electricity price was also displayed on the tablet during the experiment. A post-experiment survey indicates that almost all households in the treatment used their cellular phones, personal computers, or tablets to confirm the price to be applied to electricity consumption.

3.2.2   CR Experiments

CRs ask households to voluntarily reduce their peak-time consumption on critical days. No pecuniary incentives were provided to the households receiving CRs. CRs stimulate an intrinsic motivation to conserve energy, by which households seek to reduce their electricity consumption because of altruism or the "warm glow" (DellaVigna et al. 2012). Thus, a conservation request in this experiment is closely related to electric utilities' "emergency" appeals for conservation during peak hours (Holladay et al. 2015), and public appeals to voluntarily conserve energy through media campaigns (Reiss and White 2008).

In the first and second experiments, households assigned to the CR group were asked to reduce electricity consumption during the peak hours on critical peak days. This request was made to the households at approximately 8 p.m. the day before each critical peak day by e-mail, which they access on their cellular phones and personal computers. The notice of the request was also displayed on the tablets. Before the third and last experiments, this group received HERs. In the third and last experiments, the group did not receive CRs.

3.2.3   Sample Construction and Randomization Tests

To solicit household participation in the experiment, the Keihanna Eco-City Promotion Council sent a letter to all households in Kyotanabe City, Kizugawa City, and Seika Town, located in the south of Kyoto, Japan, in January 2012. In the letter, the Council asked these households whether they had any interest in the experiment, any access to the Internet, and owned an on-site generation facility, such as a rooftop photovoltaic. Of the 39,166 households in the experimental site, 1,649 (4.2 %) replied showing a willingness to participate in the experiment.

After eliminating the households without access to the Internet, those with on-site generation facilities, and the students living alone, the Council randomly selected 900 households and assigned them to each group in February 2012. Of the 900 households asked to participate in the experiment, 714 (79.3 %) had agreed to join the experiment by the end of March 2012, and smart meters that measured half-hourly electricity consumption were installed free of charge at their homes by the end of June 2012. The 186 remaining households refused to participate in the experiment.

The subsequent analysis estimates the average treatment effect on the treated groups, which can be estimated by the difference in the expected outcomes between the treated households (i.e., CPP, CR, and CPP + HER groups) and untreated households (control group) if the treatment is randomized across households (Wooldridge 2002). Table 3.3 summarizes the ratio of participating households across the control and three treatment groups. The ratio of participating households did not vary across these groups, which ranged from 77.6 to 80.4 %. A relatively

Table 3.3 The ratio of participating households

	Control group	CR group	CPP group	CPP + HER group	Total
Number of households asked to participate (A)	200	200	250	250	900
Number of households agreeing to participate (B)	159	160	194	201	714
B/A (%)	79.5	80.0	77.6	80.4	79.3

Source The Keihanna Eco-City Next-Generation Energy and Social Systems Demonstration Project Promotion Council

small variation in the rate across the four groups implies that no selectivity in compliance occurred in the experiment.

In fact, as shown by Table 3.4, there had been no statistically significant difference in electricity consumption and in the ratio of households' all-electricity contracts between the control and treatment groups prior to the first experiment. Prior to the experiment, the monthly data on household consumption in June 2012, and data on households' electricity contracts were obtained from the Keihanna Eco-City Promotion Council. In addition, according to the Council's survey on households' demographic and housing characteristics, and appliance holding before commencing the first experiment, there was little difference in household size, income, number of room air conditioners, floor space, and age of home across the groups. However, as shown in Table 3.5, a survey on the Kansai region [population approximately 21 million as of October 1, 2014 (JSB 2016)] indicates that the participating households lived with more members than the average household, and were more likely to live in a relatively new detached house that was spacious and equipped with a larger number of air conditioners. Thus, the response of households in the experiment to the treatment may differ from that of the average household in the Kansai region because of "site selection bias" (Allcott 2015).

To examine whether experimental randomness holds, electricity consumption prior to the experiment is regressed on a set of dummy variables for the treatment groups (Delmas and Lessem 2014). Table 3.6 indicates that none of the dummy variables was statistically significant at the 5 % level and F statistic was insignificant. These results support experimental randomness.

Experimental randomness is also examined by a probit model for being assigned to the treatment (Houde et al. 2013). Table 3.7 presents the estimation results of the probit model, in which the dependent variable takes a value of one if households were assigned to the treatment group and 0 otherwise. Treatments that did not belong to the pair were excluded in the probit model for that pair of groups. Electricity usage in June 2012 and the dummy for the all-electric contracts are employed as explanatory variables. The maximum likelihood estimates of the model indicate that no explanatory variable is statistically significant at the 5 % level, and the likelihood ratio statistic is insignificant for any pair of the control and treatment groups. These results provide evidence of randomization.

Table 3.4 Summary statistics of households' characteristics prior to the experiment

	Control group	CR group		CPP group		CPP + HER group	
	Mean	Mean	Difference	Mean	Difference	Mean	Difference
All electric = 1	0.338 (0.475)	0.389 (0.489)	0.050 (0.836)	0.310 (0.464)	−0.029 (0.521)	0.267 (0.440)	−0.071 (1.323)
Electricity usage in June 2012 (kWh/day)	12.938 (7.158)	12.468 (7.185)	−0.470 (0.575)	11.739 (5.774)	−1.199* (1.681)	12.104 (6.589)	−0.834 (1.119)
Household size	3.378 (1.181)	3.187 (1.089)	−0.191 (1.329)	3.237 (1.197)	−0.141 (1.013)	3.167 (1.232)	−0.211 (1.493)
Number of room air conditioners	3.885 (1.513)	3.669 (1.755)	−0.216 (1.018)	3.462 (1.472)	−0.424** (2.384)	3.698 (1.533)	−0.187 (1.035)
Income < $30,000	0.085 (0.282)	0.063 (0.244)	−0.023 (0.661)	0.099 (0.300)	0.013 (0.379)	0.121 (0.329)	0.036 (0.962)
Income ≥ $30,000 and < $60,000	0.393 (0.500)	0.455 (0.517)	0.062 (0.924)	0.401 (0.499)	0.008 (0.133)	0.459 (0.516)	0.065 (1.057)
Income ≥ $60,000 and < $90,000	0.274 (0.453)	0.286 (0.461)	0.012 (0.202)	0.278 (0.453)	0.004 (0.078)	0.236 (0.432)	−0.038 (0.698)
Income ≥ $90,000	0.248 (0.438)	0.196 (0.403)	−0.051 (0.925)	0.222 (0.420)	−0.026 (0.491)	0.185 (0.389)	−0.063 (1.238)
Floor space, sq. ft. < 861	0.126 (0.335)	0.171 (0.380)	0.045 (0.961)	0.227 (0.423)	0.101** (2.235)	0.214 (0.415)	0.088* (1.949)
Floor space, sq. ft. ≥ 861 and <1292	0.731 (0.490)	0.692 (0.496)	−0.039 (0.605)	0.699 (0.481)	−0.032 (0.541)	0.667 (0.507)	−0.064 (1.069)
Floor space, sq. ft. ≥ 1,292	0.143 (0.354)	0.137 (0.347)	−0.006 (0.134)	0.074 (0.262)	−0.069* (1.804)	0.119 (0.327)	−0.023 (0.563)
Age of home < 10 years	0.411 (0.494)	0.352 (0.480)	−0.059 (0.964)	0.394 (0.491)	−0.017 (0.291)	0.355 (0.480)	−0.056 (0.989)
Age of home ≥ 10 years and <20 years	0.349 (0.478)	0.368 (0.484)	0.019 (0.317)	0.388 (0.490)	0.039 (0.698)	0.424 (0.496)	0.076 (1.335)
Age of home ≥ 20 years	0.240 (0.429)	0.280 (0.451)	0.040 (0.718)	0.218 (0.414)	−0.023 (0.459)	0.221 (0.416)	−0.019 (0.393)
Detached (1 = yes)	0.884 (0.322)	0.856 (0.353)	−0.028 (0.654)	0.690 (0.464)	−0.194*** (4.266)	0.734 (0.443)	−0.150*** (3.399)

Note: The column "Difference" indicates the difference in each variable for treatment groups with respect to the control group. In the column "Mean," standard deviations are in parentheses. In the column "Difference," absolute values of *t*-statistics are in parentheses

Source The Keihanna Eco-City Next-Generation Energy and Social Systems Demonstration Project Promotion Council

*Significant at the 10 % level

**Significant at the 5 % level

***Significant at the 1 % level

3.2.4   Time-of-Day Electricity Consumption

The data on time-of-day household electricity consumption were obtained from the Keihanna Eco-City Promotion Council. Household's electricity consumption was

Table 3.5 Summary statistics of households in the Kansai region

	Kansai region
Household size	2.67
Number of room air conditioners	3.00
Income < $30,000	0.21
Income ≥ $30,000 and < $60,000	0.41
Income ≥ $60,000 and < $90,000	0.23
Income ≥ $90,000	0.15
Floor space, sq. ft. < 861	0.49
Floor space, sq. ft. ≥ 861 and < 1,292	0.27
Floor space, sq. ft. ≥ 1,292	0.24
Age of home < 10 years	0.28
Age of home ≥ 10 years and < 20 years	0.30
Age of home ≥ 20 years	0.42
Detached (1 = yes)	0.55

Notes The data on the Kansai region were obtained from a survey in which 9,000 households responded. These households did not include any participant in the experiment
Source The Keihanna Eco-City Next-Generation Energy and Social Systems Demonstration Project Promotion Council

Table 3.6 Regression of electricity consumption before the first experiment

Variables	Coefficients (standard errors)
Constant	12.967^{***} (0.593)
Dummy for CR group (1 = yes)	−0.337 (0.845)
Dummy for CPP group (1 = yes)	-1.212^{*} (0.728)
Dummy for CPP + HER group (1 = yes)	−0.939 (0.763)
Adjusted R-squared	0.001
F statistic	1.123 (p-value = 0.339)
Number of observations	658

Notes The dependent variable is daily average consumption of electricity in June 2012. White's robust standard errors are employed. Standard errors are in parentheses
*Significant at the 10 % level
***Significant at the 1 % level

recorded every half an hour during the experiments by a smart meter installed free of charge in the households. The half-hourly observations on each household's electricity consumption were aggregated into the daily data. This was done because a few observations recorded zero consumption in the control and CR groups. Zero half-hourly consumption of electricity may have been due to the limited capacity of the data-recording device installed at each household's residence.

Table 3.8 compares the daily time-of-day electricity consumption between the control and treatment groups during each experiment. In all experiments, during the

Table 3.7 Estimation results of a probit model for assignment to the treatment group

	Control versus CR group	Control versus CPP group	Control versus CPP + HER group
Electricity usage in June 2012 (kWh/day)	−0.018 (0.013)	−0.022 (0.015)	−0.009 (0.014)
Dummy for all electric (1 = yes)	0.342[*] (0.198)	0.069 (0.201)	−0.071 (0.206)
Constant	0.113 (0.158)	0.415[**] (0.163)	0.317[**] (0.155)
Likelihood ratio statistic	3.151 (p-value = 0.207)	3.016 (p-value = 0.221)	1.658 (p-value = 0.437)
Number of observations	284	327	331

Notes The dependent variable is equal to 1 if each household is assigned to the treatment group, and 0 otherwise. Observations include the control and one of the treatment groups. Treatments that did not belong to the pair were excluded in the probit model for that pair of the groups. Standard errors are in parentheses
[*]Significant at the 10 % level
[**]Significant at the 5 % level

Table 3.8 Comparison of daily electricity usage between control and treatment groups

	Mean electricity usage (kWh/day)				Difference in mean usage with respect to control		
	Control	CR	CPP	CPP +HER	CR	CPP	CPP +HER
Peak period, first	1.57 (1.20)	1.42 (1.15)	1.33 (1.00)	1.40 (1.22)	−0.15 (0.13)	−0.24[*] (0.12)	−0.17 (0.13)
All periods, first	14.80 (7.96)	14.25 (8.06)	13.29 (6.52)	13.98 (8.19)	−0.55 (0.90)	−1.51[*] (0.79)	−0.82 (0.86)
Peak period, second	3.60 (2.84)	3.29 (2.29)	2.67 (1.95)	2.87 (2.16)	−0.31 (0.29)	−0.93[***] (0.27)	−0.73[***] (0.27)
All periods, second	26.67 (20.68)	23.77 (15.97)	20.27 (15.23)	21.71 (16.99)	−2.90 (2.07)	−6.40[***] (1.97)	−4.96[**] (2.03)
Peak period, third	1.63 (1.20)	1.55 (1.19)	1.45 (1.10)	1.50 (1.27)	−0.08 (0.13)	−0.18 (0.12)	−0.13 (0.13)
All periods, third	15.42 (7.94)	14.78 (8.36)	14.08 (6.99)	14.42 (8.30)	−0.64 (0.91)	−1.34[*] (0.81)	−1.00 (0.86)
Peak period, fourth	3.61 (2.83)	3.23 (2.18)	2.64 (1.79)	2.83 (2.20)	−0.38 (0.28)	−0.97[***] (0.26)	−0.78[***] (0.27)
All periods, fourth	26.19 (20.15)	23.24 (15.69)	19.79 (14.14)	20.67 (16.23)	−2.95 (2.02)	−6.40[**] (1.89	−5.52[***] (1.97)

Notes Standard deviations are in parentheses. The peak period is from 1 p.m. (6 p.m.) to 4 p.m. (9 p.m.) in the first and third experiments (the second and fourth experiments)
Source The Keihanna Eco-City Next-Generation Energy and Social Systems Demonstration Project Promotion Council
[*]Significant at the 10 % level
[**]Significant at the 5 % level
[***]Significant at the 1 % level

peak period, the electricity consumption of each treatment group was lower than that of the control group. In all experiments, the total electricity consumption of each treatment was also lower than that of the control. While the difference in peak electricity consumption between the control and treatment receiving CPP was statistically significant in the second and fourth experiments (in winter), that difference was insignificant in the first and third experiments (in summer). In all experiments, a statistically significant difference in electricity consumption was not found between the CR and control groups.

3.3 Estimation Results

3.3.1 LA/AIDS Model

The subsequent analysis assumes that the residential demand for electricity is conditional on households' ownership of energy appliances. Without further consideration to the issue of ownership (Matsukawa and Ito 1998), it focuses on modeling the time-of-day demand for electricity conditional on durable ownership, since each experiment lasted for a couple of months, which is relatively a short time period. The analysis also assumes weak separability, which indicates the presence of the sub-utility associated only with electricity consumption. Weak separability is a necessary and sufficient condition for allocating total electricity expenditure into time-of-use electricity expenditure that is consistent with utility maximization (Deaton and Muellbauer 1980b, p. 124). The assumption of weak separability makes the analysis consistent with a two-stage budgeting framework. Under two-stage budgeting, expenditure decisions on electricity and all other goods can be represented by a recursive structure as if the household allocated income between electricity and all other goods at the first stage, and then at the second stage, decided on the expenditure for electricity demand in each time period.

The expenditure equation for electricity demand in each time period, which is conditional upon the allocation of income between electricity and all other goods, is written as $y_i = f_i(p_p, p_o, Y)$, where y_i, p_p, p_o, and Y are the expenditure for electricity demand in period i, electricity price of the peak period, electricity price of the off-peak period, and total expenditure for electricity. Due to the presence of zero electricity consumption in the control and CR groups, half-hourly electricity usage is aggregated and the overall consumption of electricity is divided into peak and off-peak consumptions. The assumption of two time periods is convenient for measuring the impact of CPP and CRs on electricity consumption for peak hours, which is the focus of Chap. 3.

An LA/AIDS model (Deaton and Muellbauer 1980a; Winters 1984; Green and Alston 1990; Pollak and Wales 1992; Matsukawa et al. 1993b) is chosen as a specification of the electricity expenditure function for households. An LA/AIDS model, which has been widely applied in empirical studies, is flexible in the sense that there are no restrictions on substitution. Another attractive feature of an LA/AIDS model is that the underlying structures associated with an indirect utility function are known and a theoretically consistent measurement of welfare is possible. Further, an LA/AIDS model dispenses with the homotheticity assumption, which indicates that given prices, the expenditure share of electricity in each time period is constant when total expenditure of electricity changes. The homotheticity assumption is made in empirical analyses of energy demand (Aigner and Hausman 1980; Caves and Christensen 1980; Jorgenson et al. 1988; Matsukawa et al. 1993a). This assumption is unduly restrictive because the expenditure elasticity of residential demand for electricity during peak periods, which must be unity under homotheticity, was statistically different from unity in the United States (Hausman and Trimble 1984) and Japan (Matsukawa 2001).

For an LA/AIDS model, the total electricity expenditure function for household i on day t is assumed to take the following form:

$$
\begin{aligned}
\log Y_{i,t} &= \left(1 - V_{i,t}\right) \log a\left(p_{p,i,t}, p_{o,i,t}\right) + V_{i,t} \log b\left(p_{p,i,t}, p_{o,i,t}\right), \\
\log a\left(p_{p,i,t}, p_{o,i,t}\right) &= \alpha + \alpha_p \log p_{p,i,t} + \alpha_o \log p_{o,i,t} + 0.5\beta_{pp}\left(\log p_{p,i,t}\right)^2 \\
&\quad + 0.5\beta_{oo}\left(\log p_{o,i,t}\right)^2 + 0.5\left(\beta_{po} + \beta_{op}\right)\log p_{p,i,t} \log p_{o,i,t}, \\
\log b(p_{p,i,t}, p_{o,i,t}) &= \log a(p_{p,i,t}, p_{o,i,t}) + \theta p_{p,i,t}^{\theta_p} p_{o,i,t}^{\theta_o},
\end{aligned}
\tag{3.1}
$$

where

$Y_{i,t}$: total electricity expenditure of household i on day t,
$V_{i,t}$: sub-utility associated with the electricity demand of household i on day t, and
$p_{k,i,t}$: electricity price for the kth period ($k = p, o$) applied to household i on day t.

In Eq. 3.1, $a(p_{p,i,t}, p_{o,i,t})$ represents subsistence electricity expenditure while $b(p_{p,i,t}, p_{o,i,t})$ represents bliss one. This is the case since along with the utility level, $a(p_{p,i,t}, p_{o,i,t})$ decreases as $b(p_{p,i,t}, p_{o,i,t})$ increases. The list of parameters associated with these functions is: $\alpha, \alpha_p, \alpha_o, \beta_{pp}, \beta_{po}, \beta_{op}, \beta_{oo}, \theta, \theta_p,$ and θ_o.

Note that the LA/AIDS model is consistent with an exact nonlinear aggregation, which makes it possible to treat aggregate consumer behavior as if it were the outcome of the decisions of a single utility-maximizing consumer. This is because the LA/AIDS model takes a form of "price independent generalized log linearity," which makes the representative expenditure level independent of prices and dependent only on the distribution of expenditures (Deaton and Muellbauer 1980a).

For the empirical analysis, in Eq. 3.1, the expenditure share function in each time period is obtained by applying Shephard's lemma to the total electricity expenditure function by replacing $V_{i,t}$ with $Y_{i,t}$, $a(p_{p,i,t}, p_{o,i,t})$, and $b(p_{p,i,t}, p_{o,i,t})$, and imposing parameter restrictions associated with adding-up, symmetry, and linear homogeneity constraints. These parameter restrictions are: $\alpha_p + \alpha_o = 1$, $\beta_{pp} + \beta_{po} = 0$, $\beta_{po} = \beta_{op}$, $\beta_{oo} + \beta_{op} = 0$, and $\theta_p + \theta_o = 0$. The effects of the CRs and HERs, cross-sectional and temporal fixed effects, and error terms are also assumed in the expenditure share function in each time period. The cross-sectional fixed effects include the impacts of income, appliance ownership, and demographic and dwelling characteristics, while temporal fixed effects include the impact of weather conditions and day-specific effects.

Specifically, for household i on day t, the expenditure share for the peak period, $S_{p,i,t}$, is given by

$$S_{p,i,t} = \alpha_p + \alpha_1 REQ_{i,t} + \left(\beta_{pp} + \alpha_2 DH_i\right) \log\left(\frac{p_{p,i,t}}{p_{o,i,t}}\right) + \theta_p \log\left(\frac{Y_{i,t}}{P_{i,t}}\right) + \omega_t + v_i + u_{i,t}$$

(3.2)

where

$REQ_{i,t}$: dummy variable equal to 1 if household i receives a conservation request on day t,
DH_i: dummy variable equal to 1 if household i received an HER before the third or last experiment,
ω_t: time fixed effects,
v_i: household fixed effects, and
$u_{i,t}$: error term.

The denominator of $Y_{i,t}$ in Eq. 3.2, denoted by $P_{i,t}$, is assumed to be approximated by Stone's price index: $log(P_{i,t}) = S_{p,i,t} \, log(p_{p,i,t}) + (1 - S_{p,i,t})log(p_{o,i,t})$. Owing to the parameter constraints, the coefficient of the log peak price is identical to the negative of the coefficient on the log off-peak price in Eq. 3.2. Thus, instead of including both the log peak and off-peak prices, the log of the ratio of peak to off-peak price is included in estimating the peak expenditure share equation.

Given total electricity expenditure and prices, the effects of CRs and HERs on the peak electricity demand are indicated by parameters α_1 and α_2, respectively. Note that the dummy variable for HERs is assumed to interact with the electricity prices in Eq. 3.2 because this dummy variable does not vary across days and is perfectly correlated with a linear combination of cross-sectional fixed effects. Although HERs were applied to both CR and CPP + HER groups before the third and fourth experiments, the effects of HERs on electricity consumption cannot be measured for households in the CR group that faced constant electricity prices of electricity during the experiments.

The CPP's effects on each household's peak usage of electricity are indicated by the uncompensated, Marshallian *total* elasticity of time-of-day electricity demand with respect to the electricity price. This elasticity is defined by the ratio of a percent change in the time-of-day electricity demand to a percent change in the time-of-day electricity price, holding *total* expenditure but not utility constant

(Caves and Christensen 1980; Mountain and Lawson 1992). For households receiving no HERs, the uncompensated total price elasticity is given by

$$E_{i,i} = \left(\frac{\beta_{ii}}{S_i} - \theta_i - 1\right) + E_{i,Y}S_i(1 + E_e), \; i = p, o, \tag{3.3}$$

$$E_{i,j} = \left(\frac{\beta_{ij}}{S_i} - \frac{\theta_i S_j}{S_i}\right) + E_{i,Y}S_j(1 + E_e), \; i, j = p, o; \; i \neq j, \tag{3.4}$$

where $E_{i,j}$ is the total elasticity of electricity demand for the ith period with respect to the electricity price for the jth period, E_e is the elasticity of the total demand for electricity with respect to the price index of electricity, and $E_{i,Y}$ is the elasticity of the electricity demand for the ith period with respect to total electricity expenditure. The expenditure elasticity $E_{i,Y}$ is given by $(\theta_i + S_i)/S_i$. These elasticities are evaluated at the sample mean. S_i represents the expenditure share of the ith period's electricity demand at the sample mean.

The first bracketed terms on the right-hand side of Eqs. 3.3–3.4 indicate the uncompensated, Marshallian *partial* price elasticities, which measure a percent change in the time-of-day electricity demand in response to a percent change in the time-of-day electricity price, holding electricity expenditure but not sub-utility constant. Four alternative formulas for these partial price elasticities using an LA/AIDS model have appeared in the literature. Using Monte Carlo experiments, Alston et al. (1994) suggested that very good approximations are provided by elasticity expressions assuming either endogenous or constant budget shares on the right-hand side of the demand equations. They also indicated that computing the elasticity expression that assumes endogenous budget shares on the right-hand side of the demand equations might not be worth the additional effort (Alston et al. 1994, p. 354). Thus, the subsequent analysis uses the so-called "LA' formula" for price elasticities (Alston et al. 1994), which assumes constant budget shares on the right-hand side of the demand equations.

3.3.2   Conservation Effects on Residential Peak Usage of Electricity

Table 3.9 summarizes the estimation results of Eq. 3.2 for four experiments using a two-way fixed effects model. The off-peak expenditure share equation was dropped to avoid singularity because of the adding-up constraint. For the two-commodity case, estimation of the parameters from either of the two share equations will yield identical results (Caves and Christensen 1980, p. 297). Standard errors are clustered at the household level to correct for serial correlation in the electricity consumption of each household in Tables 3.9 and 3.10. The number of households decreased as the experiments proceeded because of equipment failure (missing data), installation of rooftop photovoltaics, or cancellation.

Table 3.9 Parameter estimates of the peak expenditure share in Eq. 3.2: a two-way fixed effects model

	First experiment	Second experiment	Third experiment	Fourth experiment
$\log(p_{p,i,t}/p_{o,i,t})$ (electricity price ratio)	0.121^{***} (0.001)	0.154^{***} (0.001)	0.123^{***} (0.001)	0.156^{***} (0.001)
$\log(Y_{i,t}/P_{i,t})$ (electricity expenditure)	0.030^{***} (0.001)	0.009^{***} (0.001)	0.028^{***} (0.001)	0.005^{***} (0.001)
$REQ_{i,t}$ ($REQ_{i,t} = 1$ if receiving a CR)	-0.004^{***} (0.001)	-0.007^{***} (0.002)	–	–
$DH_i \cdot \log(p_{p,i,t}/p_{o,i,t})$ ($DH_i = 1$ if receiving an HER)	–	–	0.002 (0.001)	0.009^{***} (0.001)
Constant	0.025^{***} (0.003)	0.125^{***} (0.003)	0.032^{***} (0.003)	0.139^{***} (0.003)
R-squared	0.612	0.727	0.583	0.727
Number of observations	45,492	48,914	45,771	44,992
Number of households	669	661	627	608

Notes Standard errors, which are clustered at the household level, are in parentheses. The peak period is from 1 p.m. (6 p.m.) to 4 p.m. (9 p.m.) in the first and third experiments (the second and fourth experiments)
***Significant at the 1 % level

A two-way random effects model is also estimated for the peak expenditure share as an alternative analysis of panel data in Table 3.10. As shown in Table 3.10, the Hausman test for fixed and random effects regressions favors the fixed effects model in all experiments. Thus, the subsequent analysis focuses only on the estimation results of the fixed effects model.

The ratio of peak to off-peak price of electricity was significant at the 1 % level for all experiments in both fixed and random effects models. There was little difference in the estimated coefficient of the log of the electricity price ratio between the fixed and random effects models. These results indicate that CPP persistently affected the time-of-day electricity demand during the experiments. The persistence of CPP effects is consistent with Ito et al. (2015).

Table 3.11a summarizes the *partial* elasticity of electricity demand for peak hours with respect to the peak and off-peak prices of electricity at the sample mean during each experiment by using the estimated coefficient of the log of the electricity price ratio in the fixed effects model. In addition to the uncompensated partial elasticities (i.e., $E^p_{p,p}$, and $E^p_{p,o}$), the compensated partial elasticity of electricity demand during peak hours with respect to the peak and off-peak electricity prices at the sample mean, denoted by $E^{*p}_{p,p}$ and $E^{*p}_{p,o}$, is presented in the table. This compensated elasticity is defined by the ratio of a percent change in the time-of-day electricity demand to a percent change in the time-of-day electricity price, holding

Table 3.10 Parameter estimates of the peak expenditure share in Eq. 3.2: a two-way random effects model

	First experiment	Second experiment	Third experiment	Fourth experiment
$\log(p_{p,i,t}/p_{o,i,t})$ (electricity price ratio)	0.120*** (0.003)	0.154*** (0.004)	0.122*** (0.004)	0.156*** (0.007)
$\log(Y_{i,t}/P_{i,t})$ (electricity expenditure)	0.025*** (0.003)	0.003 (0.003)	0.023*** (0.004)	−0.001 (0.003)
$REQ_{i,t}$ ($REQ_{i,t} = 1$ if receiving a CR)	−0.005* (0.003)	−0.006 (0.004)	–	–
$DH_i \cdot \log(p_{p,i,t}/p_{o,i,t})$ ($DH_i = 1$ if receiving an HER)	–	–	0.002 (0.006)	0.009 (0.009)
Constant	0.038*** (0.008)	0.143*** (0.010)	0.046*** (0.010)	0.154*** (0.009)
R-squared	0.318	0.455	0.319	0.462
Hausman test statistic (p-value)	208.41 (0.000)	245.57 (0.000)	210.93 (0.000)	179.01 (0.000)
Number of observations	45,492	48,914	45,771	44,992
Number of households	669	661	627	608

Notes Standard errors, which are clustered at the household level, are in parentheses for each coefficient. The peak period is from 1 p.m. (6 p.m.) to 4 p.m. (9 p.m.) in the first and third experiments (the second and fourth experiments)
*Significant at the 10 % level
***Significant at the 1 % level

sub-utility, but not total electricity expenditure constant. The price elasticities of off-peak electricity demand (i.e., $E_{o,o}^p$, $E_{o,p}^p$, $E_{o,o}^{*p}$, and $E_{o,p}^{*p}$) are shown in Table 3.11b.

Further, the elasticity of the peak (off-peak) electricity demand with respect to total electricity expenditure, denoted by $E_{p,Y}$ ($E_{o,Y}$), is presented in Tables 3.11a and 3.11b. The relation between the compensated and uncompensated partial elasticities of the ith period's demand with respect to the jth period's price is given by the Slutsky equation: $E_{i,j}^{*p} = E_{i,j}^p + S_j E_{i,Y}$. Thus, it holds that $E_{i,j}^{*p} = -E_{i,i}^{*p}$. All elasticities in Tables 3.11a and 3.11b were statistically significant at the 1 % level.

The peak demand for electricity is found to be a substitute for the off-peak demand for electricity, as indicated by the positive estimate of $E_{p,o}^{*p}$ and $E_{o,p}^{*p}$ for all experiments in Tables 3.11a and 3.11b. Additionally, the concavity of the electricity expenditure function is indicated by the negative estimate of $E_{p,p}^{*p}$ and $E_{o,o}^{*p}$ for all experiments. As there are only two time periods, electricity consumption during these two periods must be substitutes. The positive value of the compensated

Table 3.11a Partial price elasticities of the residential peak demand for electricity

Elasticity (peak electricity price)	First experiment	Second experiment	Third experiment	Fourth experiment
$E_{p,p}^{p}$ (45 cents/kWh)	−0.301	−0.355	−0.304	−0.355
$E_{p,p}^{p}$ (65 cents/kWh)	−0.483	−0.501	−0.483	−0.500
$E_{p,p}^{p}$ (85 cents/kWh)	−0.579	−0.580	−0.577	−0.577
$E_{p,p}^{p}$ (105 cents/kWh)	−0.638	−0.630	−0.637	−0.626
$E_{p,p}^{*p}$ (45 cents/kWh)	−0.104	−0.109	−0.106	−0.109
$E_{p,p}^{*p}$ (65 cents/kWh)	−0.231	−0.188	−0.229	−0.186
$E_{p,p}^{*p}$ (85 cents/kWh)	−0.280	−0.211	−0.277	−0.207
$E_{p,p}^{*p}$ (105 cents/kWh)	−0.299	−0.213	−0.294	−0.209
$E_{p,o}^{p}$ (45 cents/kWh)	−0.881	−0.685	−0.860	−0.666
$E_{p,o}^{p}$ (65 cents/kWh)	−0.654	−0.530	−0.640	−0.516
$E_{p,o}^{p}$ (85 cents/kWh)	−0.534	−0.446	−0.525	−0.436
$E_{p,o}^{p}$ (105 cents/kWh)	−0.459	−0.393	−0.451	−0.386
$E_{p,o}^{*p}$ (45 cents/kWh)	0.104	0.109	0.106	0.109
$E_{p,o}^{*p}$ (65 cents/kWh)	0.231	0.188	0.229	0.186
$E_{p,o}^{*p}$ (85 cents/kWh)	0.280	0.211	0.277	0.207
$E_{p,o}^{*p}$ (105 cents/kWh)	0.299	0.213	0.294	0.209
$E_{p,Y}$ (45 cents/kWh)	1.182	1.040	1.164	1.021
$E_{p,Y}$ (65 cents/kWh)	1.137	1.031	1.123	1.016
$E_{p,Y}$ (85 cents/kWh)	1.113	1.026	1.102	1.014
$E_{p,Y}$ (105 cents/kWh)	1.098	1.023	1.088	1.012

Notes $E_{i,j}^{p}$ denotes the uncompensated, Marshallian partial elasticity of electricity demand during period i with respect to the electricity price of period j at the sample mean. $E_{i,j}^{*p}$ denotes the compensated partial elasticity of electricity demand during period i with respect to the electricity price of period j at the sample mean. $E_{p,Y}$ denotes the elasticity of the peak electricity demand with respect to total electricity expenditure at the sample mean. All elasticities were statistically significant at the 1 % level

cross-price elasticities implies that this condition is satisfied in the estimated equation. These results confirm the theoretical consistency of the model.

Because of relatively small compensated cross-price elasticities $E_{p,o}^{*p}$ and $E_{o,p}^{*p}$, the uncompensated cross-price elasticities, $E_{p,o}^{p}$ and $E_{o,p}^{p}$, are dominated by the income effect in the Slutsky equation and become negative. The dominance of the income effect in the Marshallian price elasticities is consistent with the empirical evidence of previous studies (Caves and Christensen 1980; Mountain and Lawson 1992). The validity of homotheticity, which is often assumed in the previous literature, can be examined by the parameter of the overall expenditure of electricity deflated with the Stone price index. The significantly positive parameter of this variable in the peak expenditure share equation implies that homotheticity does not hold in time-of-day electricity demand, and that the expenditure share of the peak (off-peak) period increases (decreases) along with the total electricity expenditure.

Table 3.11b Partial price elasticities of the residential off-peak demand for electricity

Elasticity (peak electricity price)	First experiment	Second experiment	Third experiment	Fourth experiment
$E^p_{o,o}$ (45 cents/kWh)	−0.825	−0.788	−0.824	−0.790
$E^p_{o,o}$ (65 cents/kWh)	−0.814	−0.769	−0.813	−0.769
$E^p_{o,o}$ (85 cents/kWh)	−0.804	−0.749	−0.803	−0.749
$E^p_{o,o}$ (105 cents/kWh)	−0.795	−0.730	−0.792	−0.730
$E^{*p}_{o,o}$ (45 cents/kWh)	−0.021	−0.034	−0.022	−0.035
$E^{*p}_{o,o}$ (65 cents/kWh)	−0.066	−0.082	−0.067	−0.083
$E^{*p}_{o,o}$ (85 cents/kWh)	−0.103	−0.119	−0.104	−0.119
$E^{*p}_{o,o}$ (105 cents/kWh)	−0.134	−0.146	−0.135	−0.147
$E^p_{o,p}$ (45 cents/kWh)	−0.139	−0.199	−0.143	−0.204
$E^p_{o,p}$ (65 cents/kWh)	−0.147	−0.218	−0.151	−0.224
$E^p_{o,p}$ (85 cents/kWh)	−0.154	−0.236	−0.159	−0.243
$E^p_{o,p}$ (105 cents/kWh)	−0.162	−0.254	−0.167	−0.262
$E^{*p}_{o,p}$ (45 cents/kWh)	0.021	0.034	0.022	0.035
$E^{*p}_{o,p}$ (65 cents/kWh)	0.066	0.082	0.067	0.083
$E^{*p}_{o,p}$ (85 cents/kWh)	0.103	0.119	0.104	0.119
$E^{*p}_{o,p}$ (105 cents/kWh)	0.134	0.146	0.135	0.147
$E_{o,Y}$ (45 cents/kWh)	0.964	0.988	0.967	0.993
$E_{o,Y}$ (65 cents/kWh)	0.961	0.986	0.964	0.993
$E_{o,Y}$ (85 cents/kWh)	0.959	0.985	0.962	0.992
$E_{o,Y}$ (105 cents/kWh)	0.956	0.984	0.959	0.991

Notes $E^p_{i,j}$ denotes the uncompensated, Marshallian partial elasticity of electricity demand during period i with respect to the electricity price of period j at the sample mean. $E^{*p}_{i,j}$ denotes the compensated partial elasticity of electricity demand during period i with respect to the electricity price of period j at the sample mean. $E_{o,Y}$ denotes the elasticity of the off-peak electricity demand with respect to total electricity expenditure at the sample mean. All elasticities were statistically significant at the 1 % level

The nonlinearity of the TOD expenditure in total electricity expenditure supports the validity of the LA/AIDS model, which does not impose homotheticity.

Table 3.12 presents the uncompensated *total* elasticities of the time-of-day electricity demand with respect to the electricity price in each experiment. The computation of these total elasticities needs the estimate of E_e in Eqs. 3.3–3.4 (i.e., the price elasticity of the total demand for electricity), which is hard to obtain because of little variation in the electricity price index across households during the experiment. An error correction model of residential demand for electricity is estimated in order to obtain the estimate of E_e. The appendix describes the details of the estimation procedure. The short-run price elasticity of −0.264 in the error-correction model is used for E_e.

The uncompensated total own price elasticity of the peak demand for electricity ($E_{p,p}$) in absolute terms ranges from 0.157 to 0.389, and the absolute value of $E_{p,p}$ increases along with the peak electricity price for each experiment. Given the peak

Table 3.12 Total own price elasticities of the residential time-of-day demand for electricity

Elasticity (peak electricity price)	First experiment	Second experiment	Third experiment	Fourth experiment
$E_{p,p}$ (45 cents/kWh)	−0.157	−0.174	−0.158	−0.174
$E_{p,p}$ (65 cents/kWh)	−0.298	−0.271	−0.296	−0.269
$E_{p,p}$ (85 cents/kWh)	−0.359	−0.308	−0.355	−0.305
$E_{p,p}$ (105 cents/kWh)	−0.389	−0.323	−0.385	−0.319
$E_{o,o}$ (45 cents/kWh)	−0.233	−0.233	−0.233	−0.234
$E_{o,o}$ (65 cents/kWh)	−0.263	−0.263	−0.264	−0.264
$E_{o,o}$ (85 cents/kWh)	−0.288	−0.285	−0.288	−0.285
$E_{o,o}$ (105 cents/kWh)	−0.308	−0.301	−0.309	−0.300

Note $E_{p,p}$ ($E_{o,o}$) denotes the uncompensated, Marshallian total elasticity of peak (off-peak) electricity demand with respect to the peak (off-peak) electricity price at the sample mean

electricity price, the absolute values of $E_{p,p}$ in the first and third experiments that were implemented in the summer, exceeded those in the second and last experiments that were implemented in the winter. The uncompensated total own price elasticity of the off-peak demand ($E_{o,o}$) in absolute terms ranges from 0.233 to 0.309.

The absolute values of $E_{p,p}$ are larger than those reported in the previous empirical studies on CPP. The literature reported relatively modest estimates on the own price elasticities of peak electricity demand for households. For example, Faruqui and Sergici (2010) found that the absolute values of the own price elasticities of peak electricity demand range from 0.02 to 0.10 in 15 residential experiments of CPP that were mostly implemented in the 1990s and early 2000s in the United States. In a recent CPP experiment in Connecticut in 2011, Jessoe and Rapson (2014) found that the absolute values of the own price elasticities of the peak electricity demand were less than 0.12 for households. Using experimental data on households living in Minnesota and South Dakota in 2011, Fenrick et al. (2014) found that the absolute value of the own price elasticity of peak electricity demand was 0.14, which was equal to the estimate of a CPP experiment on Japanese households in Ito et al. (2015).

Turning to CRs, as shown by the statistically significant coefficient in Tables 3.9 and 3.10, households receiving the requests reduced the ratio of daily electricity demand during peak hours to during off-peak hours in the first experiment. In the second experiment, the energy conservation effect of CRs was significant in the fixed effects model but insignificant in the random effects model. Given electricity prices and total electricity consumption, a percent change in the peak electricity demand because of CRs is given by $r\alpha_1$, where r denotes the ratio of daily total electricity consumption to daily peak electricity consumption. This ratio, r, is evaluated at the average of the control group, and the parameter α_1 is obtained from the estimation result of the fixed effects model.

As shown in Table 3.13, CRs reduced the electricity demand for peak hours by 4.0 % during the first experiment and 5.1 % in the second experiment. These

Table 3.13 Peak-reducing effects of CRs

	First experiment	Second experiment
Reduction in the peak electricity demand (%)	4.0 %	5.1 %

peak-reducing CR estimates are lower than the conservation effects of the government's call for a voluntary reduction in the peak electricity usage of households in the Kansai regions. Because of the voluntary saving, Kansai households achieved an approximately 10 % reduction in the peak electricity consumption in the summer of 2012 (JMETI 2012) and 5 % reduction in the winter of 2012/2013 (JMETI 2013). The effects of CRs are also lower than those reported by Reiss and White (2008), who found that public appeals to conserve energy without any pecuniary incentive reduced energy use by approximately 7 % in California.

Note that HERs raised the ratio of the peak usage of electricity to the off-peak usage, as shown by Tables 3.9 and 3.10. However, the interpretation of this result needs reservation, because the effects of HERs are modeled only by an interaction between a treatment dummy variable and the electricity price ratio. The effects of HERs were further investigated in Chap. 5 in a more rigorous manner.

3.4 Discussion

This section discusses the possible reasons why CPP and CRs contributed to energy conservation in the experiments. When households face CPP in the experiments, they attempted to constrain their usage of electricity during peak hours. According to a post survey on participants in the first experiment, approximately 90 % of households facing CPP reduced the usage of their room air conditioners during peak hours in summer. These households attempted to reduce the usage of room air conditioners either by raising the temperature setting or by leaving home during peak hours on critical peak days. Most CPP households also attempted to constrain their electricity usage of refrigerators by raising the temperature setting or by reducing the frequency of opening the refrigerator. These actions for energy conservation certainly reduced electricity expenditure under CPP, which could be the monetary reward that motivated households.

The energy-conservation behavior was also found in households facing CRs. According to a post survey on participants in the first experiment, approximately 85 % of CR households reduced the usage of their room air conditioners during peak hours by raising the temperature setting or by leaving home during peak hours on critical peak days in summer. The energy-conservation behavior could enhance the CR households' utility. In response to a question about the experiment, approximately 50 % of CR households replied that they enjoyed saving electricity during the first experiment. Thus, they may have had intrinsic motivations such as altruism and the "warm glow," which enhance the households' utility through energy conservation. Note, however, that in response to the same question, approximately 70 % of CPP households replied that they also enjoyed saving electricity during the first experiment. It is not clear whether monetary rewards induced CPP households to enjoy saving electricity.

The energy-conservation effects of CPP and CRs may also depend on households' attitude concerning a secure supply of electricity. The effect of households' attitude on energy conservation is particularly important because the four experiments in Kyoto were conducted after the East Japan Great Earthquake in March 2011. The latter was immediately followed by the nuclear disaster in Fukushima, which caused serious concern about the secure supply of electricity in Japan. In fact, in response to a question on energy-conservation behavior in a pre-experimental survey, approximately 66 % of participants in the first experiment replied that they became more active in energy conservation after the East Japan Great Earthquake in March 2011. The presence of these households may have raised the impacts of CPP and CRs in the experiments.

3.5 Conclusion

The empirical investigation of four randomized field experiments in southern Kyoto, Japan, during the fiscal years 2012–13, indicated that two alternative policy instruments for energy conservation, CPP and CRs, contribute to the reduction in households' electricity usage during peak periods. The empirical analysis of households' response to CPP and CRs builds on an LA/AIDS model that is more general than the traditional CES demand model and has a structure consistent with utility maximization. The application of an LA/AIDS model to the time-of-day electricity demand of households participating in the experiments, indicates that, depending on electricity prices and seasons, the absolute values of the own price elasticity of peak electricity demand ranged from 0.157 to 0.389 at the sample average.

In comparison to the previous studies, the effects of CPP are large in terms of the price elasticities of electricity demand. CRs reduced electricity demand during peak hours by 4.0–5.1 %, but these effects were lower than the effects of the government's call for a voluntary saving of peak electricity usage on households in the Kansai region. They were also lower than the effects of public appeals to conserve energy in California.

Apart from a difference in demand modeling, data, and experimental sites between this chapter and the previous studies, a possible reason for a relatively large impact of CPP in this chapter is households' behavior of saving electricity during peak hours, such as reducing the usage of room air conditioners and refrigerators during the experiments. Another possible reason is households' attitude concerning the secure supply of electricity after the nuclear disaster in Fukushima, which prompted households to be more active in saving electricity.

Appendix: Estimation of the Elasticity of Total Electricity Demand with Respect to the Electricity Price Index

This appendix measures the price elasticities of residential demand for electricity using time-series data over the period 1975–2013 in Japan. The annual data over this period are obtained from EDMC (2015). The price elasticities are estimated from an error correction model of residential demand for electricity.

The model for residential demand for electricity is assumed to take a log-linear form:

$$\log ELC_t = \alpha + \beta \log RPE_t + \gamma \log CON_t + \varepsilon_t, \tag{A.1}$$

where

ELC_t: residential consumption of electricity in year t,
RPE_t: ratio of residential electricity price index to consumer price index in year t (2010 = 100),
CON_t: consumption in year t (2010 price), and
ε_t: error term.

The electricity price is deflated by consumer price index. The consumption variable is a proxy for households' income. In Eq. A.1, parameter β indicates the price elasticity of electricity demand while γ indicates the income elasticity of electricity demand.

In order to measure the elasticity of total electricity demand with respect to the electricity price index, denoted by E_e in Eqs. 3.3–3.4, the estimation of the model in Eq. A.1 needs to account for the possible non-stationarity of the variables. In fact, statistical tests for stationarity of these variables indicate that they are not stationary and are integrated of order 1. Table A.1 summarizes unit root tests that are often employed to determine whether each variable is stationary. Specifically, the table lists the results of the Dickey–Fuller, Phillips–Perron, and Weighted Symmetric tests (Hall and Cummins 2009). These test statistics indicate the null-hypothesis of a unit root cannot be rejected for each variable in Eq. A.1. Thus, $\log ELC_t$, $\log RPE_t$, and $\log CON_t$ are found to be non-stationary variables integrated of order 1.

Because of variables integrated of order 1, stationary linear relations, i.e., cointegrating relations, may exist among these variables. The Johansen–Juselius trace test for cointegration (Hall and Cummins 2009) is conducted to explore the

Table A.1 Unit root tests

	$\log ELC_t$	$\log RPE_t$	$\log CON_t$
Dickey–Fuller test	−2.654 (0.082)	−1.093 (0.718)	−1.720 (0.421)
Phillips–Perron test	−1.689 (0.816)	−2.345 (0.739)	−1.816 (0.802)
Weighted Symmetric test	−0.555 (0.946)	−1.217 (0.709)	0.284 (0.995)

Note p-values are in the parentheses

Table A.2 Tests for cointegrating relationships among $\log ELC_t$, $\log RPE_t$, and $\log CON_t$

Null hypothesis	Alternative hypothesis	Test statistic	Eigenvalue	p-value
$r = 0$	$r = 1$	35.15	0.449	0.020
$r \leq 1$	$r = 2$	16.64	0.296	0.075
$r \leq 2$	$r = 3$	5.76	0.169	0.137

Notes The Johansen–Juselius trace test for cointegration includes a finite-sample correction. Parameter r denotes the number of cointegrating relationships. The number of lags is set equal to 1 on the basis of Akaike Information Criterion (AIC)

possible stationary relationships among the variables. Table A.2 presents test results. The number of lags is set equal to 1 on the basis of Akaike Information Criterion (AIC). Test statistics indicate that the null hypothesis of $r = 0$ (r denotes the number of cointegrating relationships) is rejected while the other two null hypotheses cannot be rejected at the 5 % level of statistical significance. This implies that there is one cointegrating vector associated with a long-run stationary relationship among $\log ELC_t$, $\log RPE_t$, and $\log CON_t$. This cointegrating vector indicates that the long-run price elasticity (i.e., β) is -0.437 and the long-run income elasticity (i.e., γ) is 0.878.

The presence of cointegrating vectors implies that the residential electricity demand model in Eq. A.1 needs to be specified as an error-correction model. Specifically, the following autoregressive distributed lag (ARDL) model is estimated:

$$\Delta \log ELC_t = a_0 + \sum_{i=1}^{s} a_i \Delta \log ELC_{t-i} + \sum_{i=0}^{m} b_i \Delta \log RPE_{t-i} + \sum_{i=0}^{n} c_i \Delta \log CON_{t-i}$$
$$+ \lambda(\log ELC_{t-1} + 0.437 \log RPE_{t-1} - 0.878 \log CON_{t-1}) + u_t,$$

$$\text{(A.2)}$$

where Δ denotes a first difference, and u_t is an error term. The fifth term on the right-hand side of Eq. A.2 represents an error correction term, which accounts for the deviation from the long-run stationary relationship. The long-run relationship is based on the estimated cointegrating vector. The parameter λ represents an adjustment coefficient and needs to be negative. The larger λ in absolute terms becomes, the more rapidly electricity demand adjusts toward its long-run level.

Table A.3 summarizes the estimation results of Eq. A.2 over the period 1977–2013 in Japan. The estimation starts by setting the number of lags equal to 1, and then successively deleting the most insignificant parameter associated with the lagged variable. This yields the ARDL model with $a_2 = b_1 = c_1 = 0$. The error-correction term is statistically significant with an adjustment coefficient (λ) of -0.259. There is no statistical evidence that suggests the presence of residual autocorrelation or autoregressive conditional heteroscedasticity, as shown by insignificant test statistics such as Breusch–Godfrey test and Lagrange multiplier test for ARCH(1) in Table A.3.

Table A.3 Estimation results of the error-correction model in Eq. A.2

	Coefficient
a_1	-0.351^{**} (0.131)
b_0	-0.264^{***} (0.070)
c_0	0.117 (0.275)
λ	-0.259^{***} (0.057)
Constant	-0.454^{***} (0.103)
Adjusted R-squared	0.558
Breusch–Godfrey test for the first-order autocorrelation	2.416
Lagrange multiplier test for ARCH(1)	1.647

Notes The dependent variable is the first difference in the natural log of annual residential demand for electricity. Standard errors are in parentheses
**Significant at the 5 % level
***Significant at the 1 % level

The estimate of b_0 indicates that the short-run price elasticity is -0.264. The short-run price elasticity is used for E_e in Eqs. 3.3–3.4, because each experiment on CPP continued for a couple of months. In absolute terms, the short-run price elasticity is lower than the long-run counterpart (-0.437) but larger than the estimate of Danish energy demand (-0.135) in Bentzen and Engsted (1993).

References

Aigner D, Hausman J (1980) Correcting for truncation bias in the analysis of experiments in time-of-day pricing of electricity. Bell J Econ 11:131–142

Allcott H (2015) Site selection bias in program evaluation. Quart J Econ 130:1117–1165

Alston J, Foster K, Green R (1994) Estimating elasticities with the linear approximate almost ideal demand system: some Monte Carlo results. Rev Econ Stat 76:351–356

Benabou R, Tirole J (2006) Incentives and prosocial behavior. Am Econ Rev 96(5):1652–1678

Bentzen J, Engsted T (1993) Short- and long-run elasticities in energy demand: a cointegration approach. Energy Econ 15:9–16

Caves D, Christensen L (1980) Econometric analysis of residential time-of-use electricity pricing experiments. J Econometrics 14:287–306

Deaton A, Muellbauer J (1980a) An almost ideal demand system. Am Econ Rev 70:312–326

Deaton A, Muellbauer J (1980b) Economics and consumer behavior. Cambridge University Press, Cambridge

DellaVigna S, List J, Malmendier U (2012) Testing for altruism and social pressure in charitable giving. Quart J Econ 127:1–56

Delmas M, Lessem N (2014) Saving power to conserve your reputation? The effectiveness of private versus public information. J Environ Econ Manag 67:353–370

EDMC (Energy Data and Modelling Center, Institute of Energy Economics, Japan) (2015) Handbook of Japan's & world energy & economic statistics. Energy Conservation Center, Japan

Faruqui A, Sergici S (2010) Household response to dynamic pricing of electricity: a survey of 15 experiments. J Regul Econ 38:193–225

Faruqui A, Sergici S, Akaba L (2014) The impact of dynamic pricing on residential and small commercial and industrial usage: new experimental evidence from Connecticut. Energy J 35(1):137–160

Fenrick S, Getachew L, Ivanov C, Smith J (2014) Demand impact of a critical peak pricing program: opt-in and opt-out options, green attitudes and other customer characteristics. Energy J 35(3):1–24

Green R, Alston J (1990) Elasticities in AIDS model. Am J Agric Econ 72:442–445

Hall B, Cummins C (2009) TSP 5.1 user's guide. TPS International

Hausman J, Trimble J (1984) Appliance purchase and usage adaptation to a permanent time-of-day electricity rate schedule. J Econometrics 26:115–139

Holladay S, Price M, Wanamaker M (2015) The perverse impact of calling for energy conservation. J Econ Behav Organ 110:1–18

Houde S, Todd A, Sudarshan A, Flora J, Armel C (2013) Real-time feedback and electricity consumption: a field experiment assessing the potential for savings and persistence. Energy J 34(1):87–102

Ito K, Ida T, Tanaka M (2015) Persistence of moral suasion and economic incentives: field experimental evidence from energy demand. NBER Working Paper 20910

Jessoe K, Rapson D (2014) Knowledge is (less) power: experimental evidence from residential energy use. Am Econ Rev 104(4):1417–1438

JMETI (Japan Ministry of Economy, Trade and Industry) (2012) Report of a sub-committee on the review of electricity demand and supply, November (in Japanese)

JMETI (Japan Ministry of Economy, Trade and Industry) (2013) Report of a sub-committee on the review of electricity demand and supply, April (in Japanese)

Jorgenson D, Slesnick D, Stoker T (1988) Two-stage budgeting and exact aggregation. J Bus Econ Stat 6:313–325

JSB (Japan Statistics Bureau) (2016) Current population estimates as of October 1, 2014 http://www.stat.go.jp/english/data/jinsui/2.htm. Accessed March 2016

List J, Price M (2013) Handbook on experimental economics and the environment. Edward Elgar

Matsukawa I (2001) Household response to optional peak-load pricing of electricity. J Regul Econ 20(3):249–267

Matsukawa I, Fujii Y, Madono S (1993a) Price, environmental regulation, and fuel demand: econometric estimates for Japanese manufacturing industries. Energy J 14(4):37–56

Matsukawa I, Madono S, Nakashima T (1993b) An empirical analysis of Ramsey pricing in Japanese electric utilities. J Jpn Int Econ 7:256–276

Matsukawa I, Ito N (1998) Household ownership of electric room air-conditioners. Energy Econ 20:375–388

Mountain D, Lawson E (1992) A disaggregated nonhomothetic modeling of responsiveness to residential time-of-use electricity rates. Int Econ Rev 33:181–207

NEPC (New Energy Promotion Council) (2016) Japan smart city portal. http://jscp.nepc.or.jp/en. Accessed January 2016

Pollak R, Wales T (1992) Demand system specification and estimation. Oxford University Press, Oxford

Reiss P, White M (2008) What changes energy consumption? Prices and public pressures. RAND J Econ 39:636–663

Winters A (1984) Separability and the specification of foreign trade functions. J Int Econ 17:239–263

Wolak F (2011) Do residential customers respond to hourly prices? Evidence from a dynamic pricing experiment. Am Econ Rev Pap Proc 101(3):83–87

Wooldridge J (2002) Econometric analysis of cross section and panel data, The MIT Press, Cambridge

Chapter 4
Effects of In-home Displays on Residential Electricity Consumption

Abstract This chapter investigates how acquiring information from in-home displays (IHDs) affects electricity usage through attention and learning, using the experimental data on the frequency of consumers' use of IHDs in summer 2012 and winter 2012/2013. Households in the treatment group could see a graph of their half-hourly electricity consumption in real time with IHDs at any time during the experiment. The immediate effect of IHDs is heightened household attention to information on electricity consumption, and the repetition of attention is expected to improve households' capacity to process information in the long run. The estimation results of the daily time-of-day electricity consumption model indicate statistically significant and persistent effects of IHD use on residential electricity consumption. The increase in IHDs' effects along with households' experience of using IHDs implies that households' capacity to process information could be improved by the repetition of attention to electricity information. Contrary to the energy-conservation literature, IHD usage was found to consistently increase residential electricity consumption because of a boomerang effect. However, an interactive effect of providing IHDs and critical peak pricing implies that providing an IHD together with pecuniary incentive schemes could be effective in energy conservation.

Keywords Real-time information feedback · Energy conservation · Inattention · Limited Information-processing capacity · Boomerang effects

4.1 Introduction

The analysis of consumer electricity consumption in response to critical peak pricing (CPP) in Chap. 3 assumes a utility-maximizing behavior of households, which implies that consumers always choose the best alternatives associated with consumption subject to income constraints. The optimal consumption choice is a basic assumption in the standard model of consumer behavior. Under this assumption, consumers perfectly recognize and attain their optimal choice of goods

© The Author(s) 2016   45
I. Matsukawa, *Consumer Energy Conservation Behavior After Fukushima*,
SpringerBriefs in Economics, DOI 10.1007/978-981-10-1097-2_4

and services by making full use of available information about price, quantity, quality, and other product attributes.

In recent years, however, a growing body of literature casts doubt on the optimal choice assumption and detects deviations from full optimization in consumer choice (DellaVigna 2009). Inattention and limited information-processing capacity are among the key factors in such consumption deviations. Inattention could lead to suboptimal consumption choices regardless of the individual's information-processing capacity (DellaVigna 2009). Individuals with limited capacity could make mistakes in choosing their optimal consumption even when fully informed (de Palma et al. 1994). Under these circumstances, deviations from full consumption optimization reduce social welfare, making policy interventions necessary to increase attention and information-processing capacity.

An extensive literature has recently investigated the effects of inattention. For example, posting price tags reduced the consumption of products sold at a grocery store because it raised the salience of commodity taxes by heightening the visibility of tax-inclusive prices (Chetty et al. 2009). Increased attention paid to electricity bills (Gilbert and Graff Zivin 2013), a nonlinear electricity pricing schedule (Kahn and Wolak 2013), and an in-home display (Houde et al. 2013; Jessoe and Rapson 2014) also reduced households' electricity consumption. On the contrary, because of lowered attention, the application of automated payment technology to car drivers reduced their response to road tolls (Finkelstein 2009). Reduced attention via the provision of automatic bill payments to households also raised their electricity consumption (Sexton 2015).

In the long run, the repetition of attention may improve the capacity to process information. A small literature provides empirical evidence on how this "learning through attention" affects behavior. Students' biased beliefs about their peak demand for cellular calling were corrected by learning through attention about their past usage (Grubb 2015; Grubb and Osborne 2015). On the contrary, learning through attention failed to improve the productive efficiency of seaweed farmers (Hanna et al. 2014) or to increase the take-up of taxpayers who received alerts about income tax credit eligibility (Manoli and Turner 2014).

The empirical investigation of CPP impacts on consumer' usage of electricity by Jessoe and Rapson (2014) is perhaps the first attempt to measure how increased attention and information-processing capacity affected the energy-conservation effects of CPP. Focusing on the policy interventions that correct for consumption biases associated with inattention and limited capacity by promoting salience and learning, Jessoe and Rapson (2014) found that households' learning through their use of in-home displays (IHDs), which enables households to obtain information about their electricity usage in real time, enhanced the energy-saving effect of electricity pricing during the peak period.

This chapter extended the work of Jessoe and Rapson (2014) by explicitly investigating how information acquisition corrects for biased beliefs, which has not been analyzed in the empirical literature on CPP. Information acquisition produces a more precise mapping of how consumption alternatives affect outcomes in the decision-making process (Simon 1955; Gabaix et al. 2006). Information acquisition is crucial for estimating salience and learning effects because inattention and limited information-processing capacity depend on how much consumers acquire available information. All else being equal, the more information consumers acquire, the more attention and capacity they achieve.

This chapter measures the effects of information acquisition on electricity consumption using data on how frequently households used IHDs in a randomized field experiment. The households could consult a graph of their half-hourly electricity consumption in real time on a tablet display at any time during the experiment. Their IHD use was automatically recorded by electronic devices installed on the premises. It is difficult to ensure that households using many electric appliances with different consumption levels attain a satisfactory level of electricity consumption. Access to consumption information in the graph enables households to raise their attention and information-processing capacity. The welfare impact of the policy intervention that provides households with IHDs depends on how much information households acquire by using their IHDs and could be measured as the incremental benefit reaped from the choice of a more satisfactory level of electricity consumption.

The estimation results of a two-way fixed effects model for the daily electricity consumption of 785 households over 68 days in the first experiment provide evidence on the energy-using effects of consumption salience: IHD use raised daily electricity consumption. The energy-using effects of consumption salience were persistent, as they were also found in the second experiment. In fact, IHD use in the second experiment increased electricity consumption more than it did in the first, implying that learning through attention improved information-processing capacity by allowing repeated attention to electricity information. The empirical evidence in this chapter provides an important policy implication: providing households with IHDs could adversely affect their energy conservation. The finding of this chapter thus contrasts with the earlier finding on IHDs' energy-saving effects on household electricity consumption (Sexton et al. 1989; Matsukawa 2004; Abrahamse et al. 2005; Darby 2006; Houde et al. 2013).

The effects of salience and learning associated with IHD use could be enhanced through additional policy interventions if providing IHDs to households is used to complement them (Price 2015). The experiment applied two additional policy instruments to households: CPP and conservation requests (CRs). A comparison among the effects of alternative interventions provides policymakers with valuable information on the effective choice of energy conservation strategies. The estimation results of the electricity consumption model in the first experiment indicate that IHD use raised the energy-saving effect of electricity pricing but did not affect the energy-saving effect of the conservation request. The interactive effect of IHDs and electricity pricing, which is consistent with Jessoe and Rapson (2014), was found to be persistent in the second experiment.

Apart from the electricity industry, this chapter's model of consumption salience and learning with information acquisition could be applied to industries where each household contracts with a single supplier for the use of multiple units of products, such as the gas, water, telecommunications, and Internet services sectors. In these industries, consumers often use multiple units of appliances, such as heating and cooking for gas, the kitchen and bath for water, cellular phones for telecommunications, and tablet computers for Internet services. Providing information about consumption assists consumers in closing the gap between their actual and optimal consumption in these industries, and the effects of salience and learning are expected to correct for biased beliefs about consumption.

The rest of this chapter proceeds as follows. Section 4.2 presents a conceptual framework outlining how information acquisition affects consumption through salience and learning. Section 4.3 explains the design of the experiments on IHD provision. Section 4.3 also describes the data obtained from the experiments and the evidence on the randomization of the experiments and then discusses stylized facts on information acquisition. Section 4.4 presents an econometric model of electricity consumption, the estimation results of the model, and the policy implications of the empirical findings. Section 4.5 concludes this chapter. The appendix provides additional estimation results.

4.2 Learning Through Attention in the Presence of Biased Beliefs About Consumption

A basic assumption of this chapter is that households' biased beliefs about consumption cause their choice of consumption to deviate from their utility-maximizing consumption. This assumption is in line with Grubb and Osborne (2015) and Allcott et al. (2014). Given income and prices that are fully known to households, the consumption of some good, denoted by x, is assumed to be given by

$$x = x^* \exp(\mu_0 + \mu_1 I), \tag{4.1}$$

where x^* is the optimal consumption of the good, I is how much information on consumption each household acquires, μ_0 is a parameter associated with the household's beliefs about consumption, and μ_1 is a parameter associated with the household's information-processing capacity. Households that are fully informed have no bias ($\mu_0 = 0$) and do not need to acquire information ($I = 0$). Thus, households achieve the optimal consumption as in the standard model of consumer behavior. Without information acquisition, however, biased beliefs about consumption lead to either over-consumption (i.e., $x > x^*$) or under-consumption (i.e., $x < x^*$) depending on the sign of μ_0: over-consumption (under-consumption) occurs if $\mu_0 > 0$ ($\mu_0 < 0$).

To see how information acquisition affects consumption, consider households that allocate their income, denoted by M, into consumption of two goods, denoted by x and z. The prices of these goods are denoted by p_x and p_z. Households' knowledge about consumption of x is assumed to be incomplete and their attention

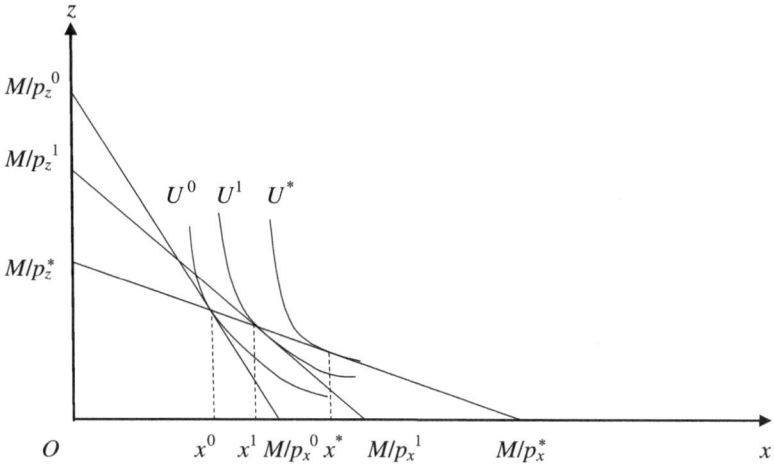

Fig. 4.1 Effects of information acquisition on consumption: $\mu_0 < 0$

to information about consumption of x is assumed to close a gap between the optimal and actual levels of consumption.

Figure 4.1 illustrates the case where biased beliefs about consumption of x are negative, that is, $\mu_0 < 0$. Given income and the actual prices of the two goods, denoted by p_x^* and p_z^*, the intercept and slope of the budget line are M/p_z^* and $-p_x^*/p_z^*$, respectively. The optimal consumption of x is x^* in the figure. The optimal consumption for households receiving a CR is to maximize utility that may include a moral cost associated with consumption (Levitt and List 2007; Ferraro and Price 2013). Without any policy intervention that provides information on consumption, the biased beliefs would result in x^0, which is lower than the optimal consumption in the figure. The difference in the two indifference curves, U^0 and U^*, reflects a welfare loss because of incomplete information on consumption.

With a policy intervention that provides information, households would acquire information on consumption of x. If $\mu_0 < 0$, for instance, information acquisition raises consumption of x to x^1 as illustrated in Fig. 4.1. The difference in the two indifference curves, U^1 and U^0, reflects a welfare improvement because of information acquisition. Whether information acquisition raises consumption of x depends on the sign of μ_0. As illustrated in Fig. 4.2, if $\mu_0 > 0$, then households' acquisition of information on consumption of x reduces consumption of x to x^1.

Given income, there exist prices of the two goods, denoted by p_x^0 and p_z^0, which yield a hypothetical budget line tangential to U^0. The intercept and slope of this hypothetical budget line are M/p_z^0 and $-p_x^0/p_z^0$, respectively, in Figs. 4.1 and 4.2. Similarly, there also exist prices of the two goods, denoted by p_x^1 and p_z^1, which yield a hypothetical budget line tangential to U^1. The intercept and slope of this hypothetical budget line are M/p_z^1 and $-p_x^1/p_z^1$, respectively, in Figs. 4.1 and 4.2.

The dual problem of utility maximization clarifies what these hypothetical budget lines indicate (Darrough and Southy 1977). Given the optimal consumption

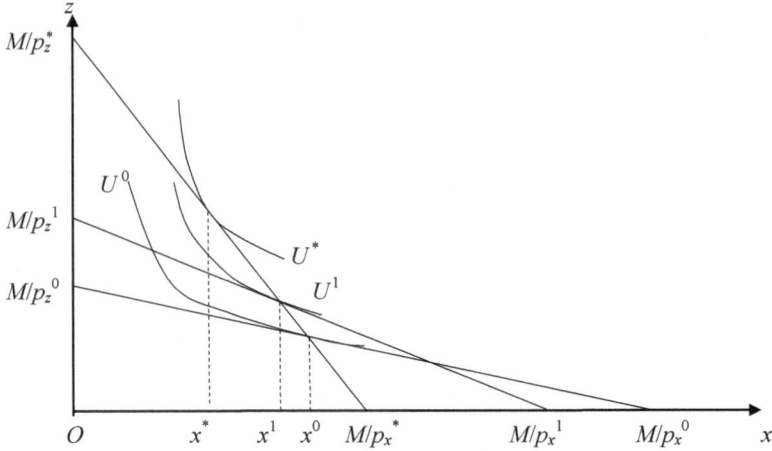

Fig. 4.2 Effects of information acquisition on consumption: $\mu_0 > 0$

and income, the solution to the dual problem of utility maximization is the set of prices minimizing utility, subject to a budget line whose intercept and slope are defined by the optimal consumption and income on the p_x–p_z plane, as illustrated by Fig. 4.3 for the case that $\mu_0 < 0$. In Fig. 4.3, each budget line whose slope is $-x^*/z^*$, $-x^0/z^0$, or $-x^1/z^1$, passes through point A, which represents the combination of p_x^* and p_z^*. This is because x^*, x^0, and x^1 are on the same budget line of $M = p_x^* x + p_z^* z$. At A, the budget line whose intercept and slope are M/z^* and $-x^*/z^*$, respectively, is tangential to an indifference curve, denoted by V^*. V^* consists of sets of prices that yield the identical level of utility, and this level of utility could be

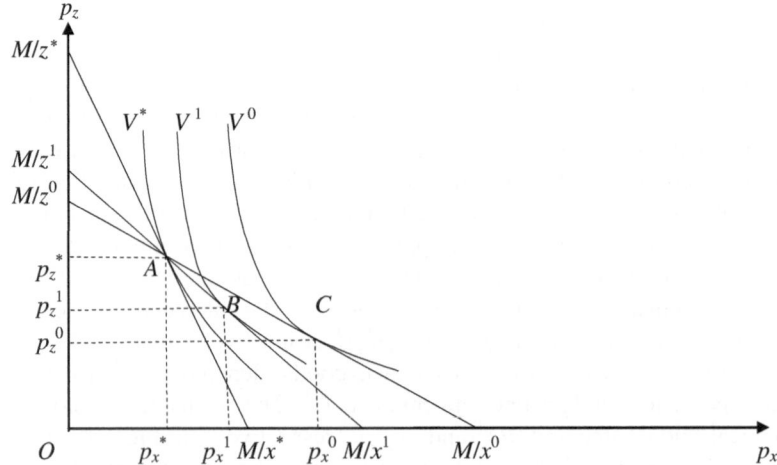

Fig. 4.3 Effects of information acquisition in a dual problem when $\mu_0 < 0$

achieved by consuming x^* and z^*. At point B, which represents the combination of p_x^1 and p_z^1, a budget line whose intercept and slope are M/z^1 and $-x^1/z^1$ is tangential to V^1. Similarly, a budget line whose intercept and slope are M/z^0 and $-x^0/z^0$ is tangential to V^0 at point C, which represents the combination of p_x^0 and p_z^0.

In the short run, policy intervention that increases households' attention to their consumption through information provision enables households to lessen their deviation from optimal consumption. If biased beliefs lead to over-consumption (under-consumption), then increasing attention through information acquisition reduces (raises) consumption (i.e., $\mu_1 < 0$ [$\mu_1 > 0$]). Thus, if $\mu_0 > 0$ ($\mu_0 < 0$), then $\mu_1 < 0$ ($\mu_1 > 0$). How much information households obtain depends on their capacity to process information. All else being equal, if households are endowed with a higher capacity to process information, the absolute value of μ_1 increases, and less information on consumption is necessary to reduce the given deviation from the optimal consumption. By contrast, households with a smaller μ_1 in absolute terms would need more information on consumption to fill the gap between actual and optimal consumption.

In the long run, policy intervention is expected to improve households' capacity to process information because households could learn about how to reduce the difference between actual and optimal consumption from their experience of adjusting consumption. The empirical literature on resource conservation, for example, finds that the long-run effects of providing households with information on conservation are persistent (Ferraro and Price 2013; Allcott and Rogers 2014). All else being equal, this long-run effect of learning through attention, indicated by an upward shift in the absolute value of μ_1, is expected to reduce households' information acquisition. Thus, the long-run effect of learning through attention could be distinguished from the short-run effect of attention alone, as in Manoli and Turner (2014).

As indicated by Hanna et al. (2014), however, learning may not help reduce the deviation from the optimal consumption because persistence in limited attention prevents households from correctly learning how to adjust their consumption. If this is so, providing households with consumption information would not improve their capacity to process information in the long run, and learning through attention would fail to fill the gap between actual and optimal consumption. Whether learning through attention is effective could be empirically examined by measuring the long-run shift in μ_1. Note that the parameter μ_0 is assumed to be constant during each experiment, because μ_0, which is included in the households fixed effects on electricity usage, cannot be identified in the empirical model of electricity consumption in Sect. 4.4.1.

Information acquisition could also depend on such costs as time and effort, as indicated by Gabaix et al. (2006). A variety of models for rational inattention have been investigated by the literature (Sims 2003; Caplin and Dean 2015; Gabaix 2015). Rational inattention with information costs implies that more salient information has lower acquisition costs. The model of this chapter does not address rational inattention because every experiment participant incurred zero cost or few costs associated with the ownership and use of an IHD.

The literature on salience provides empirical evidence that individuals are inattentive to prices and pecuniary incentives (Chetty et al. 2009; Finkelstein 2009; Gilbert and Graff Zivin 2013; Kahn and Wolak 2013; Sexton 2015). Although electricity consumption is a homogeneous good and a simple linear form of electricity pricing was applied to households during the experiment, the assumption of full attention to prices may not hold for a treatment group. This is because electricity prices that depended on dates and times were applied to households allocated to one of the treatment groups. These households received notification of electricity prices through e-mail and IHDs during the experiment. Thus, the use of IHDs may have raised attention to electricity prices. This effect of price salience will be discussed in Sect. 4.4.4.

In the subsequent analysis, the variable I in Eq. 4.1 is assumed to be associated only with the information about electricity usage households could obtain from using IHDs. Consumers' acquisition of the information about monthly electricity bills may also affect their electricity consumption. Indeed, the literature provides evidence that increased (decreased) attention to electricity bills reduced (raised) residential electricity usage (Gilbert and Graff Zivin 2013; Sexton 2015). However, the literature did not investigate to what extent households actually referred to the information about their bills, because of the lack of data. In contrast, this chapter employs data on the use of IHDs, which indicates how much information households acquire. Also, monthly electricity bills make it difficult for households to realize their daily usage of electricity during peak and off-peak hours. IHDs could provide households with information about their time-of-day usage of electricity. On these grounds, this chapter does not investigate consumers' acquisition of the information about monthly electricity bills.

4.3 Experimental Design and Data

4.3.1 IHD Experiments

In the first experiment of CPP and CRs in summer 2012, from July 23 to September 28 (68 days), the Keihanna Eco-City Promotion Council provided an IHD free of charge to households assigned to one of three treatment groups: IHD only, IHD + CPP, and IHD + CR. The control group households did not have IHDs. Except for the control, each of the three treatments was identical with one of the four groups in Chap. 3, as shown by Table 4.1: the IHD only group corresponds to the "control" group in Chap. 3; the IHD + CR group corresponds to the "CR" group in Chap. 3; and the IHD + CPP group consists of the CPP group and the CPP + HER group in Chap. 3. Because of missing data, the number of households in each group in this chapter was not equal to that in the corresponding group in Chap. 3.

Table 4.1 Definition of the control and treatments in Chap. 3 and this chapter

Chapter 3	This chapter
–	Control group
Control group	IHD only group
CR group	IHD + CR group
CPP group	IHD + CPP group
CPP + HER group	

To examine whether the salience effect of providing an IHD is persistent and whether learning through attention affects electricity consumption, the second experiment on the same households of the three treatments was conducted from December 17, 2012, to February 28, 2013 (74 days). The second experiment aimed to provide evidence on the long-run effect of learning through attention in electricity consumption, measured by an upward shift in the absolute value of parameter μ_1 in Eq. 4.1. In the first and second experiments, CPP was also applied to the IHD + CPP group and CRs were also applied to the IHD + CR group. Section 3.2 in Chap. 3 describes the details of the experimental design of CPP and CRs in these experiments.

Apart from the treatment groups, 126 control group subjects were randomly selected from approximately 3000 households living in the experimental site with a smart meter and owning no facility for onsite generation. Unfortunately, no data on the compliance rate of the control group are available. Because of a difference in the process used to select households in the treatment and control groups, the compliance rate of the control group may have been different from that of each treatment group. In Sect. 4.3.2, a statistical test for randomization will be conducted to examine whether there had been any difference between the control group and each of the treatment groups before the first experiment began.

Households allocated to the IHD-only, IHD + CPP, or IHD + CR group could see a graph of their half-hourly electricity consumption in real time on a tablet display at any time during the first and second experiments. The information feedback about the half-hourly usage of electricity in this chapter is closer to that provided in real-time usage than the feedback technologies employed in previous field experiments (Sexton et al. 1989; Matsukawa 2004). However, the interval of information feedback in this chapter is a little longer than the 10 minute interval in Houde et al. (2013) and the 15 minute interval in Jessoe and Rapson (2014).

Figure 4.4 illustrates an example of the graph in the tablet display: each bar indicates the half-hourly electricity consumption on July 25, and the solid line with dots indicates the electricity consumption on the day before. The latest information on the half-hourly consumption of electricity indicates 0.27 kWh, which is displayed in the middle of the graph in Fig. 4.4. The peak period is indicated by the shaded areas. The peak period was from 1 p.m. to 4 p.m. in summer and from 6 p.m. to 9 p.m. in winter. A comparison of electricity consumption between one day and a day in the previous week could be made by touching the area labeled "Comparison with previous week" in the tablet display. The data on the half-hourly consumption of electricity were updated in real time by an electronic device

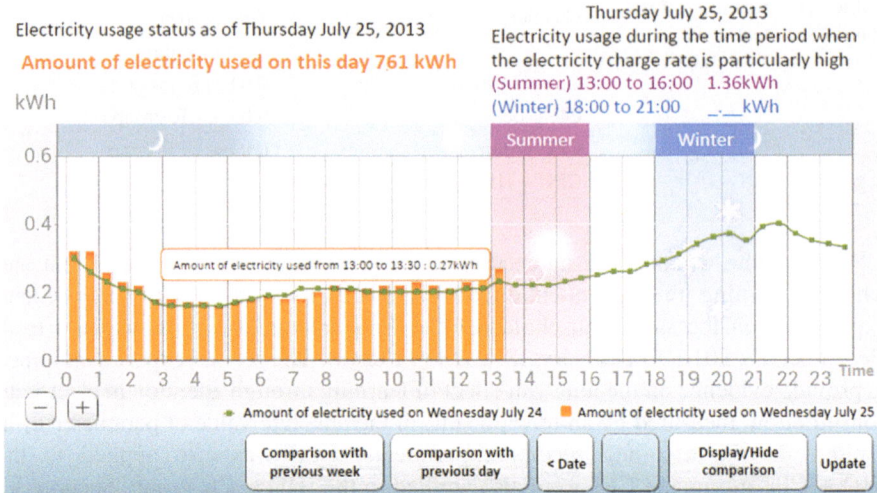

Fig. 4.4 Graph of the half-hourly electricity consumption of a household in a tablet display. *Source* Mitsubishi Heavy Industries, Ltd.

installed in each household. Households could acquire information about which appliances were most frequently used by viewing the real-time usage of electricity in the IHDs without incurring information acquisition costs. Prior to the first experiment, engineers of the Keihanna Eco-City Promotion Council visited each household to ensure that the tablet display worked properly and that the household could use it fully.

The use of IHDs is defined by how many times each household accessed to the graph illustrated in Fig. 4.4 each day. The daily access to an IHD was recorded whenever each household accessed to a graph in Fig. 4.4 by operating the IHD. Although households could see prices and conservation requests through IHDs, their access to these information items was not included in the use of IHDs.

After the first experiment, households in the three treatment groups could continue to use their IHDs every day from September 29 to December 16, 2012. Thus, before the second experiment started, households had been given sufficient time to learn about their consumption of electricity. This relatively long experience of using an IHD may promote learning through attention, which will be examined by comparing the effects of an IHD in the second experiment with those in the first experiment in Sect. 4.4.3.

4.3.2 Randomization Tests

Table 4.2 compares households' daily-average electricity consumption prior to the experiment, in June 2012, and the ratio of households' all-electricity contracts as of

Table 4.2 Households' characteristics prior to the first experiment

	Control	IHD only		IHD + CR		IHD + CPP	
	Mean	Mean	Difference	Mean	Difference	Mean	Difference
Electricity usage in June 2012 (kWh/day)	12.36 (7.35) $n = 126$	12.97 (7.06) $n = 142$	0.61 (0.69)	12.63 (7.18) $n = 142$	0.27 (0.31)	11.89 (6.20) $n = 374$	−0.47 (−0.63)
All electric = 1	0.36 (0.48) $n = 126$	0.33 (0.47) $n = 142$	−0.03 (−0.45)	0.39 (0.49) $n = 142$	0.03 (0.63)	0.28 (0.45) $n = 375$	−0.08 (−1.57)

Notes The column "Difference" indicates the difference in each variable for treatment groups with respect to the control group. In the column "Mean," standard deviations are in parentheses. In the column "Difference," t-statistics are in parentheses. The number of households is indicated by n in the table. Because of one missing observation, the number of households in the IHD + CPP group is 374 for the data on electricity consumption in June 2012
Source The Keihanna Eco-City Next-Generation Energy and Social Systems Demonstration Project Promotion Council

July 23, 2012. Unfortunately, no data on the characteristics of the control group households were obtained from a survey, which was limited to the treatment groups. No household changed electricity contracts during the experiment. Of the 659 households that received IHDs in the treatment groups, 142, denoted by "IHD + CR" in Table 4.2, received a CR on critical peak days, and 375 households, denoted by "IHD + CPP" in the table, received CPP during the experiment. The "IHD only" group consists of 142 households that received neither a CR nor CPP. Because of one missing observation, the number of households in the IHD + CPP group is 374 for the data on electricity consumption in June 2012. As shown by Table 4.2, there had been no statistically significant difference in electricity consumption between the control group and any of the treatment groups prior to the experiment, and no significant difference had been found in the ratio of households' all-electricity contracts between the control group and any of the treatment groups before the first experiment started.

As in Sect. 3.2.3 in Chap. 3, whether the experimental randomness holds is examined by regressing electricity consumption prior to the experiment on a set of dummy variables for the treatment groups and by estimating a probit model for being assigned to one of the treatment groups for each pair in the control and treatment groups. Table 4.3 indicates that none of the dummy variables was statistically significant and F statistic was not significant. Table 4.4 presents the estimation results of the probit model, in which the dependent variable takes 1 if households were assigned to the treatment group and 0 otherwise. Treatments that did not belong to the pair were excluded in the probit model for that pair of the groups. Electricity usage in June 2012 and the dummy for the all-electric contract are employed as explanatory variables. The maximum likelihood estimates of the

Table 4.3 Regression of electricity consumption before the first experiment

Variables	Coefficients (standard errors)
Constant	12.356^{***} (0.600)
Dummy for IHD only group (1 = yes)	0.611 (0.825)
Dummy for IHD + CR group (1 = yes)	0.274 (0.825)
Dummy for IHD + CPP group (1 = yes)	0.463 (0.694)
Adjusted R-squared	0.000
F statistic	1.034 (p-value = 0.377)
Number of observations	784

Notes The dependent variable is daily average consumption of electricity in June 2012. White's robust standard errors are employed. Standard errors are in parentheses
***Significant at the 1 % level

Table 4.4 Estimation results of a probit model for being assigned to the treatment group

	Control versus IHD only group	Control versus IHD + CR group	Control versus IHD + CPP group
Electricity usage in June 2012 (kWh/day)	0.014 (0.013)	−0.000 (0.012)	0.003 (0.011)
Dummy for all electric (1 = yes)	−0.181 (0.190)	0.100 (0.182)	−0.236 (0.159)
Constant	−0.036 (0.157)	0.038 (0.153)	0.704^{***} (0.129)
Likelihood ratio statistic	1.392 (p-value = 0.499)	0.394 (p-value = 0.821)	2.647 (p-value = 0.266)
Number of observations	268	268	500

Notes The dependent variable is equal to 1 if each household is assigned to the treatment group, and 0 otherwise. Treatments that did not belong to the pair were excluded in the probit model for that pair of the groups. Standard errors are in parentheses. Observations include the control and one of the treatment groups
***Significant at the 1 % level

model indicate that no explanatory variable is statistically significant, and the likelihood ratio statistic is insignificant for any pair of the control and treatment groups. These results provide evidence of randomization. Thus, the average treatment effect on the treated, which is defined as the mean effect for those who actually participated in the experiment, could be consistently estimated by the difference-in-means estimator (Wooldridge 2002).

4.3.3   *Information Acquisition and Electricity Consumption*

The data on the frequency of IHD use as well as time-of-day electricity consumption were obtained from the Keihanna Eco-City Promotion Council. The Council collected data on how many times each household used an IHD every day during the first and second experiments. These data were automatically recorded by an electronic device installed in each household.

Table 4.5 compares the daily frequency of household IHD usage during the experiment across the three treatment groups. The number of households in the

Table 4.5 Comparison of IHD usage and electricity consumption between control and treatment groups

	Control	IHD only		IHD + CR		IHD + CPP	
	Mean	Mean	Difference	Mean	Difference	Mean	Difference
IHD use, first (times/day)	–	0.51 (0.72) $n = 142$	–	0.36 (0.47) $n = 142$	–	1.75 (1.76) $n = 375$	–
IHD use, second (times/day)	–	0.29 (0.59) $n = 137$	–	0.25 (0.65) $n = 128$	–	0.90 (1.02) $n = 367$	–
Peak electricity usage, first (kWh/day)	1.72 (1.41) $n = 126$	1.57 (1.20) $n = 142$	−0.15 (−0.96)	1.43 (1.16) $n = 142$	−0.29* (−1.86)	1.37 (1.12) $n = 375$	−0.35** (−2.58)
Off-peak electricity usage, first (kWh/day)	13.11 (7.78) $n = 126$	13.18 (7.17) $n = 142$	0.07 (0.08)	12.95 (7.35) $n = 142$	−0.16 (−0.17)	12.26 (6.67) $n = 375$	−0.85 (−1.10)
Peak electricity usage, second (kWh/day)	3.60 (2.69) $n = 126$	3.64 (2.91) $n = 137$	0.04 (0.12)	3.35 (2.20) $n = 128$	−0.25 (−0.82)	2.78 (2.06) $n = 367$	−0.82*** (−3.10)
Off-peak electricity usage, second (kWh/day)	21.32 (15.90) $n = 126$	23.21 (19.08) $n = 137$	1.89 (0.87)	20.81 (14.58) $n = 128$	−0.51 (−0.27)	18.31 (14.82) $n = 367$	−3.01* (−1.85)

Notes The column "Difference" indicates the difference in each variable for treatment groups with respect to the control group. In the column "Mean," standard deviations are in parentheses. In the column "Difference," *t*-statistics are in parentheses. The number of households is indicated by *n* in the table. The peak period was from 1 p.m. to 4 p.m. in summer and from 6 p.m. to 9 p.m. in winter. The number of households decreased by 4.1 % in the second experiment, because of equipment failure (missing data), the installation of roof-top photovoltaics, or cancellation
Source The Keihanna Eco-City Next-Generation Energy and Social Systems Demonstration Project Promotion Council
*Significant at the 10 % level
**Significant at the 5 % level
***Significant at the 1 % level

treatment groups decreased from 659 in the first experiment to 632 in the second experiment because of equipment failure, the installation of roof-top photovoltaics, or cancellation. A couple of important stylized facts on information acquisition emerge from the data on the daily frequency of household IHD usage. First, of the three treatment groups, the CPP households used IHDs most during the experiment. The difference in usage between the CPP and other treatment groups persists in the first and second experiments, as illustrated by Figs. 4.5 and 4.6. The consistently higher IHD usage in the CPP group may imply the presence of an interactive effect of CPP and the use of an IHD, which will be examined in Sect. 4.4.4. The effects of CPP on IHD use will be estimated in Sect. 4.4.1.

Second, the daily frequency of household IHD usage declined in the second experiment for all treatment groups, perhaps implying that households need less information on electricity consumption in the second experiment than in the first experiment because they have improved their capacity to process information. This is consistent with the long-run effect of learning through attention discussed in Sect. 4.2. The decreased IHD usage in the second experiment may also imply, however, that households reduced their attention to consumption information because they got tired of looking at graphs on their tablet displays. If this is so, information acquisition would not affect electricity consumption in the second experiment. The effect of information acquisition on electricity consumption in the second experiment will be examined in Sect. 4.4.3.

Daily time-of-day electricity consumption is also compared between the control group and any of the treatment groups in Table 4.5. During the peak period in both experiments, the electricity consumption of the IHD + CPP group was lower than that of the control group, and the difference in peak electricity consumption

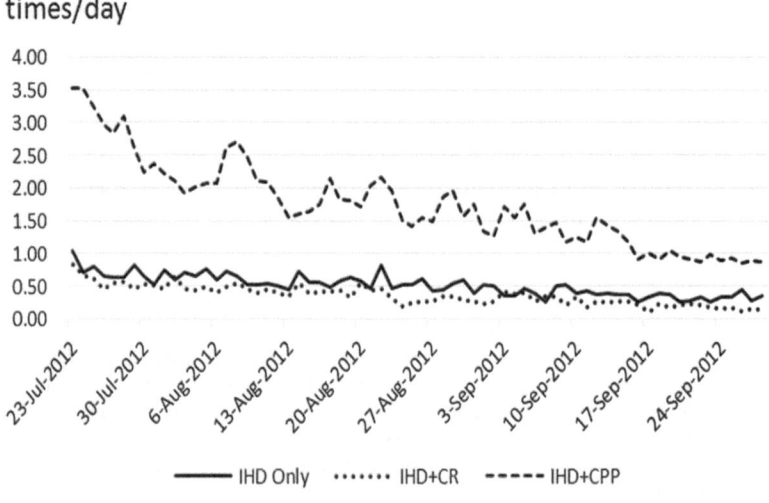

Fig. 4.5 Daily frequency of IHD usage in the first experiment. *Source* The Keihanna Eco-City Next-Generation Energy and Social Systems Demonstration Project Promotion Council

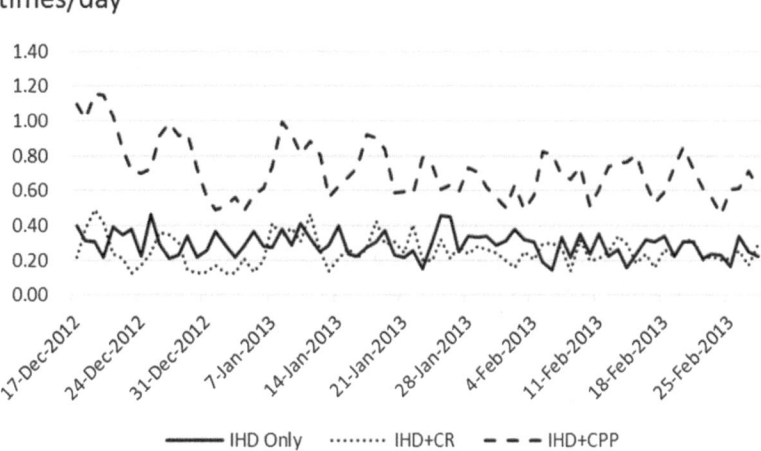

Fig. 4.6 Daily frequency of IHD usage in the second experiment. *Source* The Keihanna Eco-City Next-Generation Energy and Social Systems Demonstration Project Promotion Council

between these two groups was statistically significant. The IHD + CR group used slightly less electricity than the control group during the peak period in both experiments. The difference in peak electricity consumption between these two groups was, however, statistically significant only at the 10 % level in the first experiment and was insignificant in the second experiment. A statistically significant difference in electricity consumption was not found between the IHD-only group and the control group.

4.4 Estimation Results

4.4.1 Econometric Model of Time-of-Day Electricity Consumption

Because utility-maximization may fail in the presence of households' inattention and limited capacity for processing information, it is difficult to apply a structure model such as an LA/AIDS model in Chap. 3, which is derived from utility maximization under a two-stage budgeting framework. Instead of a structure model, a reduced-form model of household demand for time-of-day electricity is used to estimate the effects of IHD usage on electricity consumption. Specifically, the natural log of daily time-of-day electricity consumption is assumed to be a linear function of the daily frequency of IHD usage, a set of dummy variables for treatment, the fixed effects of temporal and cross-sectional factors, and unobserved errors:

$$\log KWH_{s,i,t} = c_0 + \alpha_1 IHD_{i,t} + \sum_{k=1}^{4} \beta_k D_{k,i,t} + \gamma_1 REQ_{i,t} + \omega_t + v_i + u_{i,t}, \quad (4.2)$$

where

$KWH_{s,i,t}$ electricity consumption of household i during period s (peak or off-peak) on day t, in kilowatt-hours (kWh),

$IHD_{i,t}$ frequency of household i's use of an IHD on day t,

$D_{k,i,t}$ dummy variable equal to 1 if the kth electricity price is applied to household i on day t,

$REQ_{i,t}$ dummy variable equal to 1 if household i receives a CR on day t,

ω_t time fixed effects,

v_i household fixed effects, and

$u_{i,t}$ error term.

As in Chap. 3, the daily electricity consumption model in Eq. 4.2 is estimated for the peak and off-peak periods, respectively, because the effects on peak electricity consumption are of primary concern for CPP and CR policies.

The effect of IHD usage on electricity consumption is indicated by coefficient α_1 in Eq. 4.2. Salience through the use of an IHD raises electricity consumption when $\alpha_1 > 0$. The positive effect of IHD usage, which corresponds to $\mu_1 > 0$ in Eq. 4.1, implies that biased beliefs lower households' actual consumption of electricity relative to optimal consumption (i.e., $\mu_0 < 0$ in Eq. 4.1). Contrariwise, salience through the use of an IHD reduces electricity consumption when $\alpha_1 < 0$. The adverse effect of IHD usage, corresponding to $\mu_1 < 0$ in Eq. 4.1, implies that biased beliefs increase households' actual consumption of electricity relative to optimal consumption (i.e., $\mu_0 > 0$ in Eq. 4.1).

The long-run effect of learning through attention on electricity consumption can be determined by comparing α_1 in the first experiment with that in the second. As discussed in Sect. 4.2, the improvement in information-processing capacity is expected to increase the absolute value of α_1 in the second experiment over that in the first. Note that biased beliefs about electricity consumption, μ_0, are included in household fixed effects, v_i, and that μ_0 cannot be identified in Eq. 4.2. Note also that, in Eq. 4.2, the variable $IHD_{i,t}$ is not transformed into the natural log because it could be zero.

Dummy variables associated with CPP and CRs indicate the observed exogenous factors affecting optimal electricity consumption. Each CPP dummy variable, $D_{1,i,t}$, $D_{2,i,t}$, $D_{3,i,t}$, or $D_{4,i,t}$, is equal to 1 only when an electricity price of 45, 65, 85, or 105 cents/kWh was applied to household i on day t. The dummy variable for CRs, $REQ_{i,t}$, is equal to 1 only when household i receives a CR on day t. The coefficients of these dummy variables, β_1, β_2, β_3, β_4, and γ_1, are expected to be negative because of the energy-conservation effects of CPP and CRs. Note that, for the off-peak period, dummy variables associated with CPP and CRs are included in

Eq. 4.2 to examine whether energy-saving activities undertaken in response to CPP or CRs extend beyond the peak period.

A two-way fixed effects model is assumed to account for temporal and cross-sectional effects on household electricity consumption. A two-way random effects model is also estimated, generating results close to those of a two-way fixed effects model, as shown in Table A.1 in the appendix. Note that the Hausman test for the fixed and random effects regressions cannot be conducted because of non-positive definite matrix associated with the test statistic (Greene 2007, E11-36). The time fixed effects, ω_t, include weather conditions' effects on electricity consumption. The household fixed effects, v_i, include those of biased beliefs (μ_0), demographic and housing characteristics, income, and appliance ownership. Standard errors are clustered at the household level to correct for serial correlation in the electricity consumption of each household.

Because of the likely correlation between the error term and $IHD_{i,t}$, the model in Eq. 4.2 could not be consistently estimated by least squares. In fact, ordinary least squares estimates on α_1 exceeded those in Sects. 4.4.2 and 4.4.3 by approximately 100–600 %, as shown in Table A.2 in the appendix. To correct for these biases, a predicted value of $IHD_{i,t}$ is used as an instrument for $IHD_{i,t}$ in Eq. 4.2. This predicted value is obtained from estimating a two-way fixed effects model in Eq. 4.3:

$$IHD_{i,t} = \tau_0 IHD_{i,t-1} + \sum_{k=1}^{4} \tau_k D_{k,i,t} + \tau_5 REQ_{i,t} + \kappa_t + \zeta_i + \varepsilon_{i,t}, \qquad (4.3)$$

where κ_t is time fixed effects, ζ_i is household fixed effects, $\varepsilon_{i,t}$ is the error term, and τ_k ($k = 0, 1, \ldots, 5$) is a parameter. The lagged dependent variable is added to the list of exogenous variables in Eq. 4.3.

Because of a correlation between $IHD_{i,t-1}$ and ζ_i in Eq. 4.3, least squares lead to biased estimates in Eq. 4.3. A first-differenced generalized method of moments (GMM) estimator (Arellano and Bond 1991) is applied to the estimation of Eq. 4.3 to correct for these biases. To eliminate ζ_i, taking first differences in Eq. 4.3 yields

$$\Delta IHD_{i,t} = \tau_0 \Delta IHD_{i,t-1} + \sum_{k=1}^{4} \tau_k \Delta D_{k,i,t} + \tau_5 \Delta REQ_{i,t} + \Delta \kappa_t + \Delta \varepsilon_{i,t}. \qquad (4.4)$$

Then, as predetermined instruments for $\Delta IHD_{i,t-1}$, which is correlated with $\Delta \varepsilon_{i,t}$, the lagged levels $IHD_{i,t-s}$ ($s = 2, 3, \ldots$) are used when the first-differenced estimator is applied to the estimation of Eq. 4.4. To account for serial correlation varying by household, White's period covariance is used as a weighting matrix for the estimator.

Table 4.6 presents the results of the first-differenced GMM estimation for the $IHD_{i,t}$ equation in 4.3. As suggested by Wooldridge (2002) and Roodman (2008), only the second lag of $IHD_{i,t}$ is used as a predetermined instrument for each time period to constrain the number of moment conditions because the presence of many moment conditions may lead to a biased GMM estimator. Including these predetermined instruments together with dummy variables for each period, CPP, and CR results in 137 instruments for the first experiment and 149 instruments for the second experiment. As shown in Table A.3 in the appendix, the use of the second and third lags of $IHD_{i,t}$ as predetermined instruments, leading to 202 instruments in the first experiment and 220 in the second experiment, resulted in parameter estimates close to those in Table 4.6. A system GMM estimator (Blundell and Bond 1998), an alternative to a first-differenced GMM estimator, is not used because the number of moment conditions exceeds that of the first-differenced GMM estimator and because $IHD_{i,t}$ is not highly autoregressive, as implied by the coefficient of $IHD_{i,t-1}$ that is much smaller than 1, shown in Table 4.6.

The Sargan test of over-identifying restrictions (Sargan 1958) provides mixed evidence on the validity of the instruments: the test statistic shown in Table 4.6 is insignificant in the second experiment but significant at the 10 % level in the first experiment. Note that the Sargan test requires homoscedasticity, which is not likely

Table 4.6 Estimation results of the IHD Eq. 4.3 with the first-differenced generalized method of moments (GMM) estimator

	First experiment	Second experiment
$IHD_{i,t-1}$ (first lag)	0.094*** (0.008)	0.033*** (0.004)
$D_{1,i,t}$ (1 = 45 cents/kWh)	−0.032 (0.036)	−0.011 (0.018)
$D_{2,i,t}$ (1 = 65 cents/kWh)	0.385*** (0.066)	0.051 (0.030)
$D_{3,i,t}$ (1 = 85 cents/kWh)	0.294*** (0.061)	0.078*** (0.024)
$D_{4,i,t}$ (1 = 105 cents/kWh)	0.311*** (0.060)	0.112*** (0.026)
$REQ_{i,t}$ (1 = request)	0.094** (0.039)	0.055*** (0.017)
Number of observations	43,494	45,504
Number of households	659	632
Number of periods	66	72
Number of instruments	137	149
Arellano–Bond test for first-order serial correlation	−13.02*** (p-value = 0.000)	−11.36*** (p-value = 0.000)
Arellano–Bond test for second-order serial correlation	0.981 (p-value = 0.327)	0.391 (p-value = 0.696)
Sargan test for over-identification	81.66* (p-value = 0.079)	71.1 (p-value = 0.473)

Notes The dependent variable is $IHD_{i,t}$. The second lag of $IHD_{i,t}$ is used as a predetermined instrument for $\Delta IHD_{i,t-1}$ in each time period in Eq. 4.4. Period dummy variables, which are not listed in the table, are included in the estimation of the model. The second lagged variable, $IHD_{i,t-2}$, is not included on the right-hand side of Eq. 4.3, because an Arellano–Bond test cannot reject the null hypothesis of no second-order serial correlation for the disturbances in the first-differenced equation. White's period covariance is used as a weighting matrix for the GMM estimator. Standard errors are in parentheses
*Significant at the 10 % level
**Significant at the 5 % level
***Significant at the 1 % level

to hold in the experiment, as substantial variance in IHD usage was observed in Table 4.5. An alternative test for over-identification is the Hansen test (Hansen 1982), which requires a relatively small number of instruments in GMM (Roodman 2008); this may not hold in Table 4.6. Moreover, both the Sargan and Hansen tests for over-identification may provide little information on the possibility of identifying the parameters because the validity of over-identifying restrictions is neither sufficient nor necessary for the validity of the moment conditions, and the validity of the moment conditions cannot be tested (Deaton 2010; Parente and Silva 2011).

The second-lagged dependent variable, $IHD_{i,t-2}$, is not included on the right-hand side of Eq. 4.3 because an Arellano–Bond test (Arellano and Bond 1991) cannot reject the null hypothesis of no second-order serial correlation for the disturbances in Eq. 4.4. Positive coefficients of the dummy variables for electricity prices $(D_{k,i,t})$ in both the first and second experiments are consistent with the stylized fact that households in the IHD + CPP treatment used IHDs more often than did households in the other treatment groups, as discussed in Sect. 4.3.3.

4.4.2 Effects of Consumption Salience in the Initial Experiment

Table 4.7 presents the estimation results of Eq. 4.2 for the first experiment in summer 2012. This experiment lasted for 68 days, from July 23 to September 28, 2012. The first two columns of Table 4.7 summarize the parameter estimates with all 51,810 observations. The predicted value of $IHD_{i,t}$, obtained from the first-differenced GMM estimator in Sect. 4.4.1, is used as an instrument for $IHD_{i,t}$ in the estimation of Eq. 4.2.

A few households used IHDs very often during the experiment. For example, the maximum daily IHD use reached 44 times in summer and 33 times in winter, and observations whose $IHD_{i,t}$ values exceeded 9 comprised 5.5 % of all observations in summer and 0.2 % in winter. Such extremely frequent use of an IHD may cause some bias in the estimated coefficients. To avoid this outlier effect of extremely frequent IHD use, Eq. 4.2 is estimated by excluding observations with $IHD_{i,t}$ values exceeding 9. Parameter estimates of Eq. 4.2 based on these trimmed data of 51,240 observations are summarized in the last two columns of Table 4.7. As shown in Table A.4 in the appendix, results similar to those in Table 4.7 are obtained from trimming observations with $IHD_{i,t}$ values exceeding 14 or 19.

The positive coefficients of $IHD_{i,t}$ imply that salience effects through the use of an IHD raise electricity consumption, by approximately 0.2 % for the peak period and 0.5 % for the off-peak period. These results indicate that $\mu_1 > 0$ in Eq. 4.1, which implies the presence of under-consumption (i.e., $\mu_0 < 0$). Although the coefficient of IHD usage is statistically significant at the 1 % level for the off-peak period, that coefficient is not significant even at the 10 % level for the peak period. The effects of IHD usage on peak electricity consumption when using the trimmed

Table 4.7 Effects of consumption salience: estimation results of a two-way fixed effects model of time-of-day electricity consumption in the initial experiment

	All observations		Trimmed observations	
	Peak period	Off-peak period	Peak period	Off-peak period
$IHD_{i,t}$ (daily frequency of IHD use)	0.002 (0.001)	0.005*** (0.001)	0.002 (0.002)	0.009*** (0.001)
$D_{1,i,t}$ (1 = 45 cents/kWh)	−0.013 (0.015)	0.005 (0.007)	−0.013 (0.016)	0.006 (0.007)
$D_{2,i,t}$ (1 = 65 cents/kWh)	−0.131*** (0.021)	−0.009 (0.009)	−0.131*** (0.021)	−0.010 (0.009)
$D_{3,i,t}$ (1 = 85 cents/kWh)	−0.158*** (0.020)	−0.012 (0.008)	−0.159*** (0.020)	−0.014* (0.008)
$D_{4,i,t}$ (1 = 105 cents/kWh)	−0.166*** (0.020)	−0.004 (0.008)	−0.165*** (0.020)	−0.005 (0.008)
$REQ_{i,t}$ (1 = request)	−0.066*** (0.022)	−0.017** (0.007)	−0.066*** (0.022)	−0.017*** (0.007)
Adjusted R^2	0.616	0.855	0.616	0.855
Number of observations	51,810	51,810	51,240	51,240
Number of households	785	785	785	785

Notes The dependent variable is the natural log of daily electricity consumption for the peak or off-peak period. The predicted value of $IHD_{i,t}$ (daily frequency of IHD use), which is obtained from the first-differenced generalized method of moments estimator, is used as an instrument for $IHD_{i,t}$ in the estimation of the electricity consumption equation. The last two columns exclude observations with $IHD_{i,t}$ values exceeding 9. Standard errors, which are clustered at the household level, are in parentheses
*Significant at the 10 % level
**Significant at the 5 % level
***Significant at the 1 % level

data are almost the same as those found when using all observations. The trimmed data yield a slightly larger estimate for the coefficient of $IHD_{i,t}$ during the off-peak period: households' single use of IHDs per day increases their electricity consumption by approximately 0.9 %. This coefficient remains statistically significant at the 1 % level.

Turning to CPP and CRs, the dummy variables for all levels of electricity prices and CRs exhibit negative coefficients that imply energy-saving effects for the peak period. Somewhat surprisingly, these energy-saving effects, though not significant, are also found for the off-peak period, except for the price of 45 cents/kWh. For CPP, the energy-saving effects on peak-time consumption range from approximately 1.3–16.6 % and consistently increase with electricity prices. The increased effects of CPP along with electricity prices are consistent with the rise in the absolute value of the total own price elasticity of peak electricity demand along with electricity prices in Chap. 3. All dummy variables for electricity prices except for 45 cents/kWh have statistically significant impacts at the 1 % level for the peak period. The effects of CPP on both peak and off-peak consumption when using the trimmed data are almost the same as those found when using all observations.

The energy-saving effect of CRs, statistically significant at the 1 % level, represents an approximately 6.6 % reduction in peak electricity use. This reduction in electricity consumption is in line with Reiss and White (2008), who found that public appeals to conserve energy without any pecuniary incentive reduced energy use by 7 % in California. However, the energy-saving effect of CRs for peak hours in Table 4.7 exceeds that in Chap. 3 (i.e., 4.0 % reduction in peak electricity use). This may be due to the impact of IHD use on electricity consumption, which is not taken into account in Chap. 3. For the off-peak period, the energy-saving effect of CRs is statistically significant at the 5 % level and exceeds that of any CPP price level. The effects of CRs on both peak and off-peak consumption when using the trimmed data are almost the same as those found when using all observations.

4.4.3 Persistence of Consumption Salience and Learning Effects: Evidence from the Second Experiment

Table 4.8 presents the estimation results of Eq. 4.2 for the second experiment in winter. The second experiment lasted for 74 days, from December 17, 2012, to February 28, 2013. The first two columns of Table 4.8 summarize the parameter estimates with all 54,432 observations, and the last two columns present those obtained from trimming observations with $IHD_{i,t}$ values exceeding 9. As shown in Table A.5 in the appendix, results close to those seen in Table 4.8 are obtained from trimming observations with $IHD_{i,t}$ vales exceeding 14 or 19. In Table 4.8, the predicted value of $IHD_{i,t}$, obtained from the first-differenced GMM estimator in Sect. 4.4.1, is used as an instrument for $IHD_{i,t}$ in the estimation of Eq. 4.2 for the second experiment.

The positive coefficients of $IHD_{i,t}$ imply that salience through the use of an IHD raises electricity consumption for both peak and off-peak periods in the second experiment. These results indicate that $\mu_1 > 0$ in Eq. 4.1, which implies the presence of under-consumption (i.e., $\mu_0 < 0$), and that salience effects are persistent in the second experiment. Moreover, the coefficient of IHD usage in the electricity consumption model, which was not statistically significant for the peak period in the first experiment, is now statistically significant at the 5 % level for that period in the second experiment, whether the observations are trimmed or not. For both sets of observations, the coefficient of IHD usage in the electricity consumption model remains statistically significant at the 1 % level for the off-peak period in the second experiment.

The effects of salience on electricity consumption are found to increase in the second experiment. The effects of IHD use on electricity consumption become larger than in the first experiment: approximately 0.6 % for the peak period and 0.8 % for the off-peak period when all observations are used for estimation.

Similarly, when trimming the observations exhibiting extremely frequent IHD usage (more than nine times a day), the salience and learning effects of IHD use in the first experiment exceed those in the first experiment. These results imply an upward shift in the absolute value of μ_1. Thus, the presence of learning through attention improved households' information-processing capacity in the second experiment.

Persistence is also found in the energy-conservation effects of CPP for both the peak and off-peak periods in the second experiment. Whether the observations are trimmed or not, dummy variables for all levels of electricity prices have negative coefficients for both the peak and off-peak periods, as seen in Table 4.8. The energy-saving effects of CPP on the peak consumption of electricity in the second experiment exceed those in the first experiment, ranging from approximately 11.2–

Table 4.8 Persistence of the effects of consumption salience and learning: estimation results of a two-way fixed effects model of time-of-day electricity consumption in the second experiment

	All observations		Trimmed observations	
	Peak period	Off-peak period	Peak period	Off-peak period
$IHD_{i,t}$ (daily frequency of IHD use)	0.006** (0.003)	0.008*** (0.002)	0.007** (0.003)	0.010*** (0.002)
$D_{1,i,t}$ (1 = 45 cents/kWh)	−0.113*** (0.022)	−0.009 (0.011)	−0.112*** (0.022)	−0.009 (0.011)
$D_{2,i,t}$ (1 = 65 cents/kWh)	−0.185*** (0.025)	−0.009 (0.011)	−0.184*** (0.025)	−0.009 (0.011)
$D_{3,i,t}$ (1 = 85 cents/kWh)	−0.187*** (0.025)	−0.019* (0.011)	−0.187*** (0.025)	−0.019* (0.011)
$D_{4,i,t}$ (1 = 105 cents/kWh)	−0.206*** (0.026)	−0.021* (0.011)	−0.206*** (0.026)	−0.021* (0.011)
$REQ_{i,t}$ (1 = request)	−0.028 (0.020)	−0.012 (0.009)	−0.028 (0.020)	−0.012 (0.009)
Adjusted R^2	0.678	0.908	0.678	0.907
Number of observations	54,432	54,432	54,299	54,299
Number of households	756	756	756	756

Notes The dependent variable is the natural log of daily electricity consumption for the peak or off-peak period. The predicted value of $IHD_{i,t}$ (daily frequency of IHD use), which is obtained from the first-differenced generalized method of moments estimator, is used as an instrument for $IHD_{i,t}$ in the estimation of the electricity consumption equation. The last two columns exclude observations with $IHD_{i,t}$ values exceeding 9. Standard errors, which are clustered at the household level, are in parentheses. The number of households decreased from 785 in the first experiment to 756 in the second experiment because of equipment failure (missing data), the installation of roof-top photovoltaics, or cancellation
*Significant at the 10 % level
**Significant at the 5 % level
***Significant at the 1 % level

20.6 % and consistently increasing with electricity prices. All electricity prices have statistically significant impacts on the peak consumption of electricity at the 1 % level.

In contrast with the CPP, the energy-conservation effects of CRs are not persistent. Whether the observations are trimmed or not, the negative coefficient of the dummy variable for CRs, which was statistically significant for both the peak and off-peak periods in the first experiment, becomes statistically insignificant for both periods in the second experiment. The insignificant impact of CRs in the second experiment in Table 4.8, which is in contrast with the persistent impact of CRs in Chap. 3, may be due to the impact of IHD usage.

4.4.4 Heterogeneous Effects of Salience and Learning

Inattention and limited information-processing capacity depend on household characteristics and the household's environment. Thus, the salience and learning effects of IHD usage may differ across households. This subsection examines how the effects of IHD usage depended on the application of CPP and CRs during the experiment. It also examines the dependence of these effects on how much each household had consumed electricity before the first experiment began.

Table 4.9 presents the estimation results of Eq. 4.2 with interaction terms of $IHD_{i,t}$ and each of $D_{1,i,t}$, $D_{2,i,t}$, $D_{3,i,t}$, $D_{4,i,t}$, and $REQ_{i,t}$ for both the peak and off-peak periods in the first and second experiments. Table 4.9 presents the estimation results using all observations. As shown in Table A.6 in the appendix, the results with the trimmed observations are close to those found with all observations, shown in Table 4.9. Without CPP or CRs, using an IHD raises electricity consumption, an effect statistically significant at the 1 % level for both the peak and off-peak periods in the first and second experiments. For both periods, the effect of IHD use on electricity consumption in the second experiment exceeds that in the first experiment. These results are consistent with the estimation results of Eq. 4.2 without the interaction terms of $IHD_{i,t}$ and each of $D_{1,i,t}$, $D_{2,i,t}$, $D_{3,i,t}$, $D_{4,i,t}$, and $REQ_{i,t}$, shown in Tables 4.7 and 4.8.

When CPP is applied to households, the effects of IHD use on electricity consumption during the peak period turn to *energy-saving*, as indicated by the negative coefficient of each interaction term "$IHD_{i,t} \times D_{k,i,t}$," which slightly exceeds in absolute terms the coefficient of $IHD_{i,t}$ for the peak period, shown in Table 4.9. Except for the electricity price of 85 cents per kWh in the first experiment, the coefficient of each interaction term "$IHD_{i,t} \times D_{k,i,t}$" is statistically significant as either the 1 or 5 % level for the peak period in both the first and second experiments. Thus, the salience and learning effects of using an IHD on electricity consumption depend on the application of CPP.

The *energy-saving effect* of IHD usage with CPP on the peak consumption of electricity is in sharp contrast with the *energy-using effect* of IHD usage without CPP. This difference in the effects of IHDs implies that consumption salience raises

Table 4.9 Heterogeneity in salience and learning effects across electricity prices

	First experiment		Second experiment	
	Peak period	Off-peak period	Peak period	Off-peak period
$IHD_{i,t}$ (daily frequency of IHD use)	0.006^{***} (0.002)	0.007^{***} (0.001)	0.016^{***} (0.004)	0.011^{***} (0.002)
$D_{1,i,t}$ (1 = 45 cents/kWh)	−0.001 (0.015)	0.012 (0.008)	-0.099^{***} (0.023)	−0.005 (0.011)
$D_{2,i,t}$ (1 = 65 cents/kWh)	-0.116^{***} (0.021)	−0.007 (0.010)	-0.171^{***} (0.025)	−0.005 (0.012)
$D_{3,i,t}$ (1 = 85 cents/kWh)	-0.144^{***} (0.020)	−0.007 (0.009)	-0.173^{***} (0.025)	−0.014 (0.012)
$D_{4,i,t}$ (1 = 105 cents/kWh)	-0.144^{***} (0.020)	0.005 (0.009)	-0.184^{***} (0.026)	−0.015 (0.012)
$REQ_{i,t}$ (1 = request)	-0.063^{***} (0.022)	-0.017^{**} (0.007)	−0.024 (0.020)	−0.011 (0.009)
$IHD_{i,t} \times D_{1,i,t}$	-0.007^{**} (0.003)	-0.005^{***} (0.001)	-0.021^{***} (0.007)	-0.005^{*} (0.003)
$IHD_{i,t} \times D_{2,i,t}$	-0.008^{**} (0.004)	−0.002 (0.002)	-0.018^{***} (0.007)	-0.006^{*} (0.003)
$IHD_{i,t} \times D_{3,i,t}$	−0.008 (0.005)	−0.003 (0.002)	-0.019^{**} (0.009)	-0.006^{*} (0.003)
$IHD_{i,t} \times D_{4,i,t}$	-0.012^{***} (0.004)	-0.005^{***} (0.002)	-0.029^{***} (0.009)	-0.008^{**} (0.003)
$IHD_{i,t} \times REQ_{i,t}$	−0.009 (0.009)	−0.001 (0.004)	−0.016 (0.012)	−0.003 (0.005)
Adjusted R^2	0.616	0.855	0.679	0.908
Number of observations	51,810	51,810	54,432	54,432
Number of households	785	785	756	756

Notes The dependent variable is the natural log of daily electricity consumption for the peak or off-peak period. The predicted value, which is obtained from the first-differenced generalized method of moments estimator, is used as an instrument for $IHD_{i,t}$ in the estimation of the electricity consumption equation. Standard errors, which are clustered at the household level, are in parentheses. The number of households decreased from 785 in the first experiment to 756 in the second experiment because of equipment failure (missing data), the installation of roof-top photovoltaics, or cancellation
*Significant at the 10 % level
**Significant at the 5 % level
***Significant at the 1 % level

the peak consumption of electricity and price salience lowers it. Without CPP, households would not have attended to electricity prices that were constant during the experiment, and consumption salience dominates price salience in the effects of IHD usage. When applying CPP, however, IHD use would have increased households' attention to prices as well as consumption because CPP stimulates the extrinsic motivation to conserve electricity. For households using IHDs, the energy-saving effect of increased attention to prices slightly exceeds the energy-using effect of increased attention to consumption. Thus, as indicated by Jessoe and Rapson (2014), providing IHDs to households enhanced CPP's energy-saving effects.

The energy-saving effect of price salience can be illustrated by Fig. 4.2. Because of biased beliefs about prices, for instance, households that consider p_x^0 and p_z^0 to be the true prices would choose x^0 as the solution to their utility maximization subject to income constraints. This exceeds the optimal consumption x^*. Households' acquisition of information on prices through using IHDs could decrease consumption to x^1 because information acquisition makes them consider p_x^1 and p_z^1, which are closer to the actual prices p_x^* and p_z^*, to be the true prices. This adverse effect of price salience on consumption is also found in the previous literature (Chetty et al. 2009; Gilbert and Graff Zivin 2013; Kahn and Wolak 2013).

For the off-peak period, the energy-saving effect of IHD usage along with CPP is not found in the first or second experiment, as the coefficient of each interaction

Table 4.10 Heterogeneity in salience and learning effects across electricity usage

	First experiment		Second experiment	
	Peak period	Off-peak period	Peak period	Off-peak period
$IHD_{i,t}$ (daily frequency of IHD use)	0.009 (0.008)	0.013*** (0.005)	0.028** (0.012)	0.012 (0.011)
$IHD_{i,t} \times \log(KWH\_6_i)$	−0.003 (0.003)	−0.003* (0.002)	−0.009** (0.005)	−0.002 (0.004)
$D_{1,i,t}$ (1 = 45 cents/kWh)	−0.013 (0.016)	0.005 (0.007)	−0.113*** (0.022)	−0.009 (0.011)
$D_{2,i,t}$ (1 = 65 cents/kWh)	−0.131*** (0.021)	−0.009 (0.009)	−0.185*** (0.025)	−0.009 (0.011)
$D_{3,i,t}$ (1 = 85 cents/kWh)	−0.158*** (0.020)	−0.011 (0.008)	−0.186*** (0.025)	−0.019* (0.011)
$D_{4,i,t}$ (1 = 105 cents/kWh)	−0.166*** (0.020)	−0.003 (0.008)	−0.206*** (0.026)	−0.021* (0.011)
$REQ_{i,t}$ (1 = request)	−0.066*** (0.022)	−0.017** (0.007)	−0.028 (0.020)	−0.012 (0.009)
Adjusted R^2	0.616	0.855	0.678	0.908
Number of observations	51,744	51,744	54,288	54,288
Number of households	784	784	754	754

Notes The dependent variable is the natural log of daily electricity consumption for the peak or off-peak period. The variable KWH_6_i indicates daily average electricity consumption in June 2012, prior to the first experiment. The predicted value of $IHD_{i,t}$ (daily frequency of IHD use), which is obtained from the first-differenced generalized method of moments estimator, is used as an instrument for $IHD_{i,t}$ in the estimation of the electricity consumption equation. Standard errors, which are clustered at the household level, are in parentheses. Because of missing observations on electricity consumption in June 2012, the number of households decreased from 785 to 784 in the first experiment and from 756 to 754 in the second experiment
*Significant at the 10 % level
**Significant at the 5 % level
***Significant at the 1 % level

term "$IHD_{i,t} \times D_{k,i,t}$" is, in absolute terms, lower than the coefficient of $IHD_{i,t}$. The dependence of the effects of IHD use on CRs is not found, as the coefficient of the interaction term "$IHD_{i,t} \times REQ_{i,t}$" is not statistically significant in either the first or second experiment for any time period.

Table 4.10 summarizes the estimation results of Eq. 4.2 with an interaction term of $IHD_{i,t}$ and the natural log of household i's daily average electricity consumption in June 2012, denoted by KWH_6_i. Because of missing electricity consumption values for June 2012, the number of households decreased from 785 to 784 in the first experiment and from 756 to 754 in the second experiment. For both the peak and off-peak periods, the coefficient of the interaction term "$IHD_{i,t} \times \log(KWH\_6_i)$" is negative in both experiments, implying that the salience and learning effects of IHD usage on electricity consumption decrease along with the level of electricity consumption. However, the empirical evidence on the dependence of salience and learning on electricity usage is weak: the coefficient of the interaction term is statistically significant only at the 10 % level for the off-peak period and is insignificant for the peak period in the first experiment. In the second experiment, the coefficient of the interaction term is statistically significant at the 5 % level for the peak period but is insignificant for the off-peak period.

4.5 Discussion

The finding on the energy-using effect of households' IHD usage contrasts with the previous finding on the energy-conservation effect of the presence of an IHD (Sexton et al. 1989; Matsukawa 2004; Abrahamse et al. 2005; Darby 2006; Houde et al. 2013). A couple of explanations for this difference in an IHD's effect on electricity consumption are possible. First, this study's model for household electricity consumption explicitly includes the effect of information acquisition, which has not been investigated in the literature that focuses on the presence of an IHD. For example, Houde et al. (2013) indicate that the presence of a real-time feedback technology yielded reductions in residential electricity consumption an average of 5.7 %. Houde et al. (2013), however, did not take account of significant heterogeneity in the treatment effect regarding how much households actually used IHDs. Information acquisition through IHD use should be investigated rather than the presence of IHDs because the difference in the frequency of IHD usage across households (observed in Table 4.5 in Sect. 4.3.3) affects electricity consumption.

Second, a "boomerang effect" of IHD use may occur among households that had already reduced electricity usage more than they had planned before starting IHD use (Ayres et al. 2009). In the wake of the nuclear accident in Fukushima, many have been deeply concerned about the security of Japan's electricity supply. This concern may have motivated Japanese households to save electricity. In fact, according to a survey by the Keihanna Eco-City Promotion Council, approximately

4,500 out of 9,000 households in the Kansai region undertook activities in 2012 that conserved energy in response to rising concern about the electricity supply. As a result, Kansai households achieved an approximately 10 % reduction in the peak electricity consumption in the summer of 2012 (JMETI 2012) and 5 % reduction in the winter of 2012/2013 (JMETI 2013). Without IHDs, these households could not have realized how much electricity consumption they had reduced and they may have reduced their consumption more than they had targeted. This may have led to under-consumption of electricity (i.e., $\mu_0 < 0$) prior to the experiment. Because of under-consumption, providing Kansai households with IHDs would have raised their electricity usage (i.e., $\mu_1 > 0$) by letting them be aware of excessive electricity saving and increase their consumption of electricity to achieve the conservation targets.

The possibility that respondents in the experiment had been highly motivated by rising concern about the electricity supply prior to the experiment may make it difficult to generalize the results of this chapter. However, as in the 1970s oil shocks and California's electricity crisis, markets for energy often experience shocks that require rapid changes in consumption, and how energy consumption responds to policy interventions becomes crucial (Reiss and White 2008). This chapter provides unique evidence on how residential electricity consumption responds to IHD provision when market shocks occur.

The electricity demand model in Eq. 4.2 assumes that households use IHDs first and then determine their usage of electricity to achieve the targeted usage, because the focus of this chapter is on how IHD use affects electricity consumption. This causal direction of the effect may be opposite: households determine electricity usage first and then check IHDs to see if they attain the targeted usage. For instance, households owning the larger number of electric appliances may be more likely to check IHDs while those who are out of home longer may be less likely to check IHDs. As long as these effects are constant during each experiment, they are included in households' fixed effects of Eq. 4.3. For the case that these effects depend on time, however, it is necessary to conduct a simultaneous estimation of the IHD equation and the electricity usage equation, which requires an additional set of exogenous variables.

This chapter investigates how learning through attention affects electricity consumption by comparing the effects of households' IHD use on electricity consumption between the first and second experiments. This comparison assumes that the speed of learning of households is relatively slow. However, providing households with information about their half-hourly usage of electricity in real-time may enable households to learn much more quickly. If that is the case, instead of comparing the average effects of IHD use between two experiments, the analysis of a time trend of IHD effects for each experiment may be appropriate (Matsukawa 2015). Also, a model of consumer learning through Bayesian updating (Grubb 2015) could be a promising alternative for investigating the effects of IHD use.

4.6 Conclusion

Inattention and limited information-processing capacity are key factors in deviations from full consumption optimization. Inattention could lead to suboptimal choices of consumption for individuals regardless of their capacity to process information. Individuals with limited information-processing capacity could make mistakes in choosing their optimal consumption even if they are fully informed. Deviations from full consumption optimization reduce social welfare, forcing policy interventions to increase individuals' attention and information-processing capacity.

This chapter uses panel data on how frequently each household uses an IHD in a randomized field experiment to investigate how acquiring information from an IHD affects electricity consumption through consumption salience and learning. Households in the treatment group could see a graph of their half-hourly electricity consumption in real time with IHDs at any time during the experiment. Providing an IHD is a promising policy intervention that corrects for the consumption biases associated with inattention and limited capacity by promoting salience and learning. The immediate effect of providing an IHD is heightened household attention to information on consumption, and the repetition of attention is expected to improve households' capacity to process information in the long run.

The estimation results of the electricity consumption model indicate statistically significant and persistent effects of salience through IHD use on residential electricity consumption. The increase in IHDs' effects along with households' experience of using IHDs implies that households' capacity to process information could be improved by the repetition of attention to electricity information. The empirical evidence offered by this chapter also indicates that providing an IHD raised household electricity consumption. Thus, providing households with IHDs, which was found to be an effective policy instrument for energy conservation in previous studies, could have an adverse effect on energy conservation. However, an interactive effect of IHD provision and electricity pricing implies that providing an IHD together with pecuniary incentive schemes could be effective in energy conservation.

The information IHDs provided to households was limited to total electricity use, and no information about the electricity use of each appliance was provided in the experiments. As in Attari et al. (2014) and Chen et al. (2014), if data on electricity use by appliance are available, the effects of information acquisition could be measured for each appliance. This chapter assumes that households' ownership of energy appliances is fixed and does not address how information provision affects households' appliance choices. The empirical literature on energy efficiency has recently investigated the effects of information provision on consumer choices of energy appliances (Sallee 2013; Davis and Metcalf 2014; Houde 2014; Yeomans and Herberich 2014; Allcott and Taubinsky 2015). Future research in this area could examine the effects of IHDs on the choice of energy appliances and appliances' electricity consumption using a discrete–continuous model of households' ownership and use of each energy appliance (Hausman 1979; Dubin and McFadden 1984; Matsukawa 2012).

Appendix: Additional Estimation Results

See Tables A.1, A.2, A.3, A.4, A.5 and A.6.

Table A.1 Effects of consumption salience: estimation results of a two-way random effects model of time-of-day electricity consumption

	First experiment		Second experiment	
	Peak period	Off-peak period	Peak period	Off-peak period
$IHD_{i,t}$ (daily frequency of IHD use)	0.002 (0.001)	0.005*** (0.001)	0.005** (0.002)	0.008*** (0.002)
Constant	0.136*** (0.025)	2.388*** (0.020)	0.917*** (0.027)	2.720*** (0.028)
$D_{1,i,t}$ (1 = 45 cents/kWh)	−0.022 (0.015)	0.003 (0.007)	−0.113*** (0.022)	−0.011 (0.011)
$D_{2,i,t}$ (1 = 65 cents/kWh)	−0.135*** (0.021)	−0.009 (0.009)	−0.178*** (0.024)	−0.009 (0.011)
$D_{3,i,t}$ (1 = 85 cents/kWh)	−0.161*** (0.019)	−0.012 (0.008)	−0.182*** (0.025)	−0.020* (0.011)
$D_{4,i,t}$ (1 = 105 cents/kWh)	−0.171*** (0.020)	−0.005 (0.008)	−0.202*** (0.026)	−0.022* (0.011)
$REQ_{i,t}$ (1 = request)	−0.066*** (0.022)	−0.016** (0.006)	−0.022 (0.020)	−0.011 (0.009)
Adjusted R^2	0.004	0.002	0.007	0.001
Number of observations	51,810	51,810	54,432	54,432
Number of households	785	785	756	756

Notes The dependent variable is the natural log of daily electricity consumption for the peak or off-peak period. The predicted value of $IHD_{i,t}$ (daily frequency of IHD use), which is obtained from the first-differenced generalized method of moments estimator, is used as an instrument for $IHD_{i,t}$ in the estimation of the electricity consumption equation. Standard errors, which are clustered at the household level, are in parentheses
*Significant at the 10 % level
**Significant at the 5 % level
***Significant at the 1 % level

Table A.2 Ordinary least squares estimates of a two-way fixed effects model of time-of-day electricity consumption

	First experiment		Second experiment	
	Peak period	Off-peak period	Peak period	Off-peak period
$IHD_{i,t}$ (daily frequency of IHD use)	0.015*** (0.002)	0.013*** (0.001)	0.019*** (0.004)	0.015*** (0.002)

(continued)

Table A.2 (continued)

	First experiment		Second experiment	
	Peak period	Off-peak period	Peak period	Off-peak period
Constant	0.119*** (0.004)	2.380*** (0.002)	0.910*** (0.007)	2.713*** (0.003)
$D_{1,i,t}$ (1 = 45 cents/kWh)	−0.011 (0.015)	0.005 (0.007)	−0.111*** (0.022)	−0.009 (0.011)
$D_{2,i,t}$ (1 = 65 cents/kWh)	−0.141*** (0.019)	−0.011 (0.008)	−0.184*** (0.024)	−0.008 (0.011)
$D_{3,i,t}$ (1 = 85 cents/kWh)	−0.165*** (0.020)	−0.015* (0.008)	−0.190*** (0.025)	−0.021* (0.011)
$D_{4,i,t}$ (1 = 105 cents/kWh)	−0.173*** (0.020)	−0.007 (0.008)	−0.211*** (0.026)	−0.023** (0.011)
$REQ_{i,t}$ (1 = request)	−0.069*** (0.021)	−0.015** (0.006)	−0.032 (0.020)	−0.013 (0.009)
Adjusted R^2	0.618	0.857	0.680	0.908
Number of observations	53,380	53,380	55,944	55,944
Number of households	785	785	756	756

Notes The dependent variable is the natural log of daily electricity consumption for the peak or off-peak period. Standard errors, which are clustered at the household level, are in parentheses
*Significant at the 10 % level
**Significant at the 5 % level
***Significant at the 1 % level

Table A.3 Estimation results of Eq. 4.3 with the first-differenced generalized method of moments (GMM) estimator: using the second and third lags of $IHD_{i,t}$ as predetermined instruments

	First experiment	Second experiment
$IHD_{i,t-1}$ (first lag)	0.116*** (0.006)	0.049*** (0.004)
$D_{1,i,t}$ (1 = 45 cents/kWh)	0.005 (0.031)	−0.014 (0.018)
$D_{2,i,t}$ (1 = 65 cents/kWh)	0.318*** (0.055)	0.057* (0.030)
$D_{3,i,t}$ (1 = 85 cents/kWh)	0.284*** (0.050)	0.073*** (0.024)
$D_{4,i,t}$ (1 = 105 cents/kWh)	0.308*** (0.053)	0.116*** (0.026)
$REQ_{i,t}$ (1 = request)	0.087** (0.035)	0.046*** (0.017)
Number of observations	43,494	45,504
Number of households	659	632
Number of periods	66	72
Number of instruments	202	220
Arellano–Bond test for first-order serial correlation	−12.82*** (*p*-value = 0.000)	−11.21*** (*p*-value = 0.000)

(continued)

Table A.3 (continued)

	First experiment	Second experiment
Arellano–Bond test for second-order serial correlation	1.756^* (p-value = 0.079)	0.987 (p-value = 0.324)
Sargan test for over-identification	185.60^{***} (p-value = 0.001)	180.89^{**} (p-value = 0.015)

Notes The dependent variable is $IHD_{i,t}$. The second and third lags of $IHD_{i,t}$ are used as predetermined instruments for $\Delta IHD_{i,t-1}$ in each time period in Eq. 4.4. Period dummy variables, which are not listed in the table, are included in the estimation of the model. The second lagged variable, $IHD_{i,t-2}$, is not included on the right-hand side of Eq. 4.3, because an Arellano–Bond test cannot reject the null hypothesis of no second-order serial correlation for the disturbances in the first-differenced equation in 4.4. White's period covariance is used as a weighting matrix for the GMM estimator. Standard errors are in parentheses
*Significant at the 10 % level
**Significant at the 5 % level
***Significant at the 1 % level

Table A.4 Estimation results of a two-way fixed effects model of time-of-day electricity consumption in the initial experiment: alternative trimming

	Trimming observations with $IHD_{i,t}$ values exceeding 14		Trimming observations with $IHD_{i,t}$ values exceeding 19	
	Peak period	Off-peak period	Peak period	Off-peak period
$IHD_{i,t}$ (daily frequency of IHD use)	0.002 (0.002)	0.007^{***} (0.001)	0.002 (0.002)	0.006^{***} (0.001)
$D_{1,i,t}$ (1 = 45 cents/kWh)	−0.012 (0.015)	0.005 (0.007)	−0.013 (0.015)	0.005 (0.007)
$D_{2,i,t}$ (1 = 65 cents/kWh)	-0.130^{***} (0.021)	−0.010 (0.009)	-0.130^{***} (0.021)	−0.010 (0.009)
$D_{3,i,t}$ (1 = 85 cents/kWh)	-0.159^{***} (0.020)	−0.013 (0.008)	-0.158^{***} (0.020)	−0.013 (0.008)
$D_{4,i,t}$ (1 = 105 cents/kWh)	-0.166^{***} (0.020)	−0.004 (0.008)	-0.167^{***} (0.020)	−0.004 (0.008)
$REQ_{i,t}$ (1 = request)	-0.066^{***} (0.022)	-0.017^{***} (0.007)	-0.066^{***} (0.022)	-0.017^{***} (0.007)
Adjusted R^2	0.615	0.855	0.616	0.855
Number of observations	51,623	51,623	51,741	51,741
Number of households	785	785	785	785

Notes The dependent variable is the natural log of daily electricity consumption for the peak or off-peak period. The predicted value of $IHD_{i,t}$ (daily frequency of IHD use), which is obtained from the first-differenced generalized method of moments estimator, is used as an instrument for $IHD_{i,t}$ in the estimation of the electricity consumption equation. The first two columns exclude observations with $IHD_{i,t}$ values exceeding 14, and the last two columns exclude observations with $IHD_{i,t}$ values exceeding 19. Standard errors, which are clustered at the household level, are in parentheses
*Significant at the 10 % level
**Significant at the 5 % level
***Significant at the 1 % level

Table A.5 Estimation results of a two-way fixed effects model of time-of-day electricity consumption in the second experiment: alternative trimming

	Trimming observations with $IHD_{i,t}$ values exceeding 14		Trimming observations with $IHD_{i,t}$ values exceeding 19	
	Peak period	Off-peak period	Peak period	Off-peak period
$IHD_{i,t}$ (daily frequency of IHD use)	0.007** (0.003)	0.009*** (0.002)	0.007** (0.003)	0.009*** (0.002)
$D_{1,i,t}$ (1 = 45 cents/kWh)	−0.113*** (0.022)	−0.009 (0.011)	−0.113*** (0.022)	−0.009 (0.011)
$D_{2,i,t}$ (1 = 65 cents/kWh)	−0.185*** (0.025)	−0.009 (0.011)	−0.185*** (0.025)	−0.009 (0.011)
$D_{3,i,t}$ (1 = 85 cents/kWh)	−0.187*** (0.025)	−0.019* (0.011)	−0.187*** (0.025)	−0.019* (0.011)
$D_{4,i,t}$ (1 = 105 cents/kWh)	−0.206*** (0.026)	−0.021* (0.011)	−0.206*** (0.026)	−0.021* (0.011)
$REQ_{i,t}$ (1 = request)	−0.029 (0.020)	−0.012 (0.009)	−0.029 (0.020)	−0.012 (0.009)
Adjusted R^2	0.678	0.908	0.678	0.908
Number of observations	54,407	54,407	54,421	54,421
Number of households	756	756	756	756

Notes The dependent variable is the natural log of daily electricity consumption for the peak or off-peak period. The predicted value of $IHD_{i,t}$ (daily frequency of IHD use), which is obtained from the first-differenced generalized method of moments estimator, is used as an instrument for $IHD_{i,t}$ in the estimation of the electricity consumption equation. The first two columns exclude observations with $IHD_{i,t}$ values exceeding 14, and the last two columns exclude observations with $IHD_{i,t}$ values exceeding 19. Standard errors, which are clustered at the household level, are in parentheses. The number of households decreased from 785 in the first experiment to 756 in the second experiment because of equipment failure (missing data), the installation of roof-top photovoltaics, or cancellation
*Significant at the 10 % level
**Significant at the 5 % level
***Significant at the 1 % level

Table A.6 Heterogeneity in salience and learning effects across electricity prices: trimming observations with $IHD_{i,t}$ values exceeding 9

	First experiment		Second experiment	
	Peak period	Off-peak period	Peak period	Off-peak period
$IHD_{i,t}$ (daily frequency of IHD use)	0.008*** (0.003)	0.011*** (0.001)	0.020*** (0.004)	0.013*** (0.002)
$D_{1,i,t}$ (1 = 45 cents/kWh)	0.002 (0.017)	0.012 (0.008)	−0.096*** (0.023)	−0.005 (0.011)

(continued)

Table A.6 (continued)

	First experiment		Second experiment	
	Peak period	Off-peak period	Peak period	Off-peak period
$D_{2,i,t}$ (1 = 65 cents/kWh)	-0.111^{***} (0.024)	-0.006 (0.011)	-0.169^{***} (0.026)	-0.005 (0.012)
$D_{3,i,t}$ (1 = 85 cents/kWh)	-0.137^{***} (0.023)	-0.004 (0.010)	-0.167^{***} (0.026)	-0.013 (0.012)
$D_{4,i,t}$ (1 = 105 cents/kWh)	-0.140^{***} (0.023)	0.004 (0.009)	-0.180^{***} (0.027)	-0.014 (0.012)
$REQ_{i,t}$ (1 = request)	-0.062^{***} (0.023)	-0.017^{**} (0.007)	-0.024 (0.020)	-0.011 (0.009)
$IHD_{i,t} \times D_{1,i,t}$	-0.011^{**} (0.004)	-0.004^{**} (0.002)	-0.025^{***} (0.008)	-0.006^{*} (0.003)
$IHD_{i,t} \times D_{2,i,t}$	-0.012^{**} (0.006)	-0.003 (0.003)	-0.022^{***} (0.009)	-0.006 (0.004)
$IHD_{i,t} \times D_{3,i,t}$	-0.013^{**} (0.006)	-0.006^{**} (0.002)	-0.029^{***} (0.010)	-0.010^{**} (0.004)
$IHD_{i,t} \times D_{4,i,t}$	-0.015^{**} (0.006)	-0.006^{***} (0.002)	-0.036^{***} (0.010)	-0.010^{***} (0.004)
$IHD_{i,t} \times REQ_{i,t}$	-0.010 (0.011)	-0.003 (0.004)	-0.020 (0.015)	-0.005 (0.005)
Adjusted R^2	0.616	0.855	0.678	0.907
Number of observations	51,240	51,240	54,299	54,299
Number of households	785	785	756	756

Notes The dependent variable is the natural log of daily electricity consumption for the peak or off-peak period. The predicted value, which is obtained from the first-differenced generalized method of moments estimator, is used as an instrument for $IHD_{i,t}$ in the estimation of the electricity consumption equation. Standard errors, which are clustered at the household level, are in parentheses. The number of households decreased from 785 in the first experiment to 756 in the second experiment because of equipment failure (missing data), the installation of roof-top photovoltaics, or cancellation
*Significant at the 10 % level
**Significant at the 5 % level
***Significant at the 1 % level

References

Abrahamse W, Steg L, Vlek C, Rothengatter T (2005) A review of intervention studies aimed at household energy conservation. J Environ Psychol 25:273–291

Allcott H, Mullainathan S, Taubinsky D (2014) Energy policy with externalities and internalities. J Public Econ 112:72–88

Allcott H, Rogers T (2014) The short-run and long-run effects of behavioral interventions: experimental evidence from energy conservation. Am Econ Rev 104(10):3003–3037

Allcott H, Taubinsky D (2015) Evaluating behaviorally motivated policy: experimental evidence from the lightbulb market. Am Econ Rev 105(8):2501–2538

Arellano M, Bond S (1991) Some tests of specification for panel data: Monte Carlo evidence and an application to employment equations. Rev Econ Stud 58:277–297

Attari S, Gowrisankaran G, Simpson T, Marx S (2014). Does information feedback from in-home devices reduce electricity use? Evidence from a field experiment. NBER Working Paper 20809

Ayres I, Raseman S, Shih A (2009) Evidence from two large field experiments that peer comparison feedback can reduce residential energy usage. NBER Working Paper 15386

Blundell R, Bond S (1998) Initial conditions and moment restrictions in dynamic panel data models. J Econ 87:115–143

Caplin A, Dean M (2015) Revealed preference, rational inattention, and costly information acquisition. Am Econ Rev 105(7):2183–2203

Chen V, Delmas M, Kaiser W, Locke S (2014) What can we learn from high frequency appliance level energy metering? Results from a field experiment. E^3 Working Paper 080

Chetty R, Looney A, Kroft K (2009) Salience and taxation: theory and evidence. Am Econ Rev 99 (4):1145–1177

Darby S (2006) The effectiveness of feedback on energy conservation: a review for DEFRA of the literature on metering, billing, and direct displays. University of Oxford, Environmental Change Institute

Darrough M, Southy C (1977) Duality in consumer theory made simple: the revealing of Roy's identity. Can J Econ 10:307–317

Davis L, Metcalf G (2014) Does better information lead to better choices? Evidence from energy-efficiency labels. E2e Working Paper 015

Deaton A (2010) Instruments, randomization, and learning about development. J Econ Lit 48:424–455

DellaVigna S (2009) Psychology and economics: evidence from the field. J Econ Lit 47:315–372

de Palma A, Myers G, Papageorgiou Y (1994) Rational choice under an imperfect ability to choose. Am Econ Rev 84:419–440

Dubin J, McFadden D (1984) An econometric analysis of residential electric appliance holdings and consumption. Econometrica 52:345–362

Ferraro P, Price M (2013) Using nonpecuniary strategies to influence behavior: evidence from a large-scale field experiment. Rev Econ Stat 95:64–73

Finkelstein A (2009) E-Z tax: tax salience and tax rates. Quart J Econ 124:969–1010

Gabaix X (2015) A sparsity-based model of bounded rationality. Quart J Econ 130:1661–1710

Gabaix X, Laibson D, Moloche G, Weinberg S (2006) Costly information acquisition: experimental analysis of a boundedly rational model. Am Econ Rev 96(4):1043–1068

Gilbert B, Graff Zivin J (2013) Dynamic salience with intermittent billing: evidence from smart electricity meters. NBER Working Paper 19510

Greene W (2007) LIMDEP Version 9.0: econometric modeling guide, Vol 1, Econometric Software Inc

Grubb M (2015) Consumer inattention and bill-shock regulation. Rev Econ Stud 82:219–257

Grubb M, Osborne M (2015) Cellular service demand: biased beliefs, learning, and bill shock. Am Econ Rev 105(1):234–271

Hanna R, Mullainathan S, Schwartzstein J (2014) Learning through noticing: theory and evidence from a field experiment. Quart J Econ 129:1311–1353

Hansen L (1982) Large sample properties of generalized method of moments estimators. Econometrica 50:1029–1054

Hausman J (1979) Individual discount rates and the purchase and utilization of energy-using durables. Bell J Econ 10:33–54

Houde S (2014) How consumers respond to environmental certification and the value of energy information. E2e Working Paper 007

Houde S, Todd A, Sudarshan A, Flora J, Armel C (2013) Real-time feedback and electricity consumption: a field experiment assessing the potential for savings and persistence. Energ J 34(1):87–102

Jessoe K, Rapson D (2014) Knowledge is (less) power: experimental evidence from residential energy use. Am Econ Rev 104(4):1417–1438

JMETI (Japan Ministry of Economy, Trade and Industry) (2012) Report of a sub-committee on the review of electricity demand and supply. November (in Japanese)

JMETI (Japan Ministry of Economy, Trade and Industry) (2013) Report of a sub-committee on the review of electricity demand and supply. April (in Japanese)

Kahn M, Wolak F (2013) Using information to improve the effectiveness of nonlinear pricing: evidence from a field experiment. Working paper

Levitt S, List J (2007) What do laboratory experiments measuring social preferences reveal about the real world? J Econ Perspect 21:153–174

Manoli D, Turner N (2014) Nudges and learning: evidence from informational interventions for low-income taxpayers. NBER Working Paper 20718

Matsukawa I (2004) The effects of information on residential demand for electricity. Energ J 25(1):1–17

Matsukawa I (2012) The welfare effects of environmental taxation on a green market where consumers emit a pollutant. Environ Resour Econ 52:87–107

Matsukawa I (2015) Information acquisition, dynamic pricing of electricity, and conservation requests: evidence from a field experiment, Mimeo. http://ssrn.com/author=1505264

Parente P, Silva S (2011) A cautionary note on tests for overidentifying restrictions. Discussion Paper 11/11, University of Exeter

Price M (2015) Using field experiments to address environmental externalities and resource scarcity: major lessons learned and new directions for future research. NBER Working Paper 20870

Reiss P, White M (2008) What changes energy consumption? Prices and public pressures. RAND J Econ 39:636–663

Roodman D (2008) A note on the theme of too many instruments. Working Paper 125, Center for Global Development

Sallee J (2013) Rational inattention and energy efficiency. NBER Working Paper 19545

Sargan J (1958) The estimation of economic relationships using instrumental variables. Econometrica 26:393–415

Sexton S (2015) Automatic bill payment and salience effects: evidence from electricity consumption. Rev Econ Stat 97:229–241

Sexton R, Sexton T, Wann J, Kling C (1989) The conservation and welfare effects of information in a time-of-day pricing experiment. Land Econ 65:272–279

Simon H (1955) A behavioral model of rational choice. Quart J Econ 69:99–118

Sims C (2003) Implications of rational inattention. J Monetary Econ 50:665–690

Wooldridge J (2002) Econometric analysis of cross section and panel data. The MIT Press

Yeomans M, Herberich D (2014) An experimental test of the effect of negative social norms on energy-efficient investments. J Econ Behav Organ 108:187–197

Chapter 5
Energy-Saving Effects of Home Energy Reports

Abstract Using data drawn from a 2013 field experiment, this chapter examines the energy-saving effects of home energy reports (HERs). HERs provide consumers with energy conservation tips and compare consumers' energy usage with that of similar neighbors. These comparisons categorized consumers as "energy-using," "average," or "energy-saving." The energy usage comparisons provide social norm information, thereby inducing consumers to conserve energy. In contrast to the HER experiments in the United States that sent the report to households by mail, the staff visited each participating household and explained the HERs in detail during the experiment. The empirical results imply that the effects of HERs depended on electricity contracts. Therefore, HERs contributed to a reduction in electricity usage, in households with all-electric contracts, while the energy-saving effects were not found in households with standard contracts. The electricity-saving effect of HERs on the all-electric households ranged from 4.0–8.7 %, which was found in each category of electricity usage. Overall, there was no indication of so-called "boomerang" effects, which raise electricity usage of households categorized as "energy-saving." The electricity-saving effect of HERs was found to become larger during the mornings and evenings of weekdays. This effect could persist over several weeks after providing HERs to some groups of households.

Keywords Peer effects · Conservation tips · Boomerang effects · All-electric contract · Targeting · Neighbor comparison

5.1 Introduction

This chapter focuses on the energy-conservation effects of home energy reports (HERs), which provide households with energy-saving tips and peer comparisons of energy usage. Energy-saving tips are expected to help households use their home appliances more efficiently. Peer comparison is expected to provide intrinsic motivation to households, often referred to as the "warm-glow" or altruism. This comparison creates more awareness of energy-saving social norms.

© The Author(s) 2016   81
I. Matsukawa, *Consumer Energy Conservation Behavior After Fukushima*,
SpringerBriefs in Economics, DOI 10.1007/978-981-10-1097-2_5

Using data from a randomized field experiment, the objective of this chapter is to investigate how HERs affect household electricity consumption. The HERs provided participating households with the neighbor comparisons and personalized electricity-saving tips in a similar way to previous studies. For the neighbor comparisons, illustrated by colored graphs in the report, each household was categorized as "energy-using," "average," or "energy-saving." The report also included electricity-saving tips that targeted different households in terms of their historical usage of electricity.

A growing body of the literature has investigated the conservation effects of HERs (Schultz et al. 2007; Nolan et al. 2008; Ayres et al. 2009; Allcott 2011, 2015; Matsukawa 2011; Costa and Kahn 2013; Ferraro and Price 2013; LaRiviere et al. 2014; Allcott and Rogers 2014; Allcott and Kessler 2015). The literature presents mixed results on the conservation effects of peer comparisons. For example, Ayres et al. (2009) indicate that reports on peer comparisons mailed monthly or quarterly contributed to the reduction in electricity and gas consumption of U.S. households; electricity saving was approximately 2.1 % on average. The conservation effects of peer comparisons were also found in residential usage of water in the U.S. (Ferraro and Price 2013). On the contrary, LaRiviere et al. (2014) and Costa and Kahn (2013) indicate that peer comparisons could raise electricity consumption of U.S. households. Allcott (2011) reported that providing energy-conservation tips to U.S. households contributed to the reduction in electricity usage by approximately 2.0 % on average, but neighbor comparisons did not result in electricity saving in the U.S. Turning to Japanese households, Arimura et al. (2013) found no significant effect of peer comparisons on energy saving.

The HERs experiment in this chapter differs from previous studies in an important way. In this study, the engineering staff of the experiment directly advised each participating household on the efficient use of energy appliances when providing an HER. The experiment lasted for approximately two months, from May 9 to July 5, 2013. During the experiment, the staff visited each participating household once and explained the details about neighbor comparisons of hourly and monthly electricity usage in August 2012. Upon visiting, the staff also explained the personalized tips for saving electricity. The direct advice would enable households to understand how to save electricity more easily than the mailed reports used in the previous studies. The advantage of face-to-face visits is that the experiment staff member could heighten the household's level of awareness regarding the neighbor comparisons and explain these comparisons. In contrast, it is difficult to heighten households' awareness using mailed reports.

The rest of this chapter proceeds as follows. Section 5.2 describes the design of the experiments and the data obtained from the experiment. Section 5.3 presents an empirical model of residential demand for half-hourly electricity consumption. This section also discusses the estimation results of the model, and the policy implications of the empirical findings. Section 5.4 concludes this chapter. The appendix provides additional estimation results.

5.2   Experimental Design and Data

Table 5.1 presents the definition of the control and treatment groups in the HER experiment. The HER experiment used the same households as those participating in the CPP and CR experiments in Chap. 3. The HER treatments consisted of Groups 2 and 4. In reference to Chap. 3, Group 2 is identical to the "CR group," while Group 4 is identical to the "CPP + HER group." The control consists of Groups 1 and 3. Group 1 is identical to the control in Chap. 3, and the "CPP group" in Chap. 3 is now Group 3. Since the test results in Chap. 3 provide evidence of randomization in the experiment, the average treatment effect of HERs on the treated, which is defined as the mean effect of HERs for households that actually participated in the experiment, could be consistently estimated by the difference-in-means estimator (Wooldridge 2002).

From May 9 to July 5 in 2013, the experiment staff visited each participating household allocated to Groups 2 or 4 once (Fig. 5.1). Using an HER, the staff members advised each household on electricity-saving tips and explained the neighbor comparisons of hourly and monthly electricity usage in August 2012. In contrast to the previous studies that sent HERs monthly or quarterly by mail (Ayres et al. 2009; Allcott 2011), households received HERs only once and they received no feedback during the experiment. However, the staff's visit to their homes would have enabled households to gain sufficient information about electricity-saving tips and neighbor comparisons.

Figure 5.2 illustrates an example of an HER. The report was printed in color on a single A4 sheet of paper. It included two key components: neighbor comparisons and personalized tips for saving electricity. The neighbor comparisons consist of three graphs: a bar chart comparing the household's hourly electricity usage averaged over the weekdays of August 2012 to the average of all participating households, a chart indicating the rank of the households in the distribution of monthly electricity usage of the treatments (Groups 2 and 4) in August 2012 (excluding weekends and holidays), and a chart indicating which category of electricity usage (i.e., "energy-using," "average," or "energy-saving") was applied to the households in Groups 2 and 4. Participating households were labeled as "energy-using" if they used more than the 33rd percentile of their neighbor comparison group, "energy-saving" if they used less than the 66th percentile of their neighbor comparison group, and "average" if they were in between those percentiles. In these categories, each household's electricity usage was compared to

Table 5.1 Definition of the control and treatments in Chaps. 3 and 5

Chapter 3	This chapter (name of the group)
Control group	Control group (Group 1)
CR group	HER group (Group 2)
CPP group	Control group (Group 3)
CPP + HER group	HER group (Group 4)

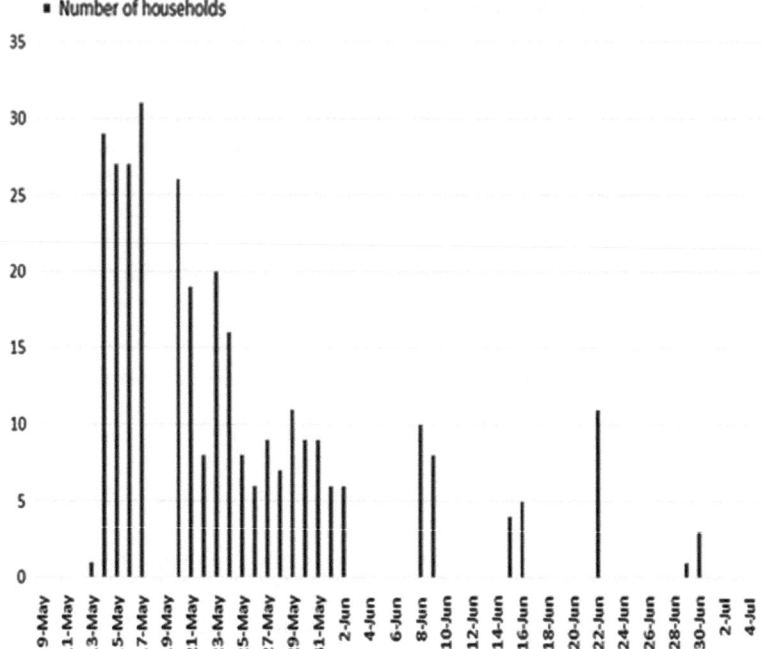

Fig. 5.1 Distribution of dates when households received HERs. *Source* The Keihanna Eco-City Next-Generation Energy and Social Systems Demonstration Project Promotion Council

households with similar demographic characteristics. Table 5.2 summarizes the distribution of electricity usage categories. HERs included limited text explaining these graphs, and the staff members explained them to each household on their visits.

The report also provided the pattern of daily electricity usage that was applied to each household. All participating households were categorized into one of seven patterns of daily electricity usage averaged over weekdays in August 2012. The report in Fig. 5.2 indicates that the household was categorized as the "elephant" type, which implies that electricity usage exceeded the average in every hour of the day during the summer.

Personalized tips for saving electricity depended on the electricity consumption type of households. For example, a report in Fig. 5.2, which suggests discontinuing the simultaneous use of two room air conditioners, using blinds or screens in the windows, and replacing an old refrigerator, was aimed at the "elephant" type households. The staff members directly explained these tips to each household, and demonstrated how these tips were effective in conserving electricity.

The data on half-hourly electricity consumption during the experiment were obtained from the Keihanna Eco-City Promotion Council. An electronic device in

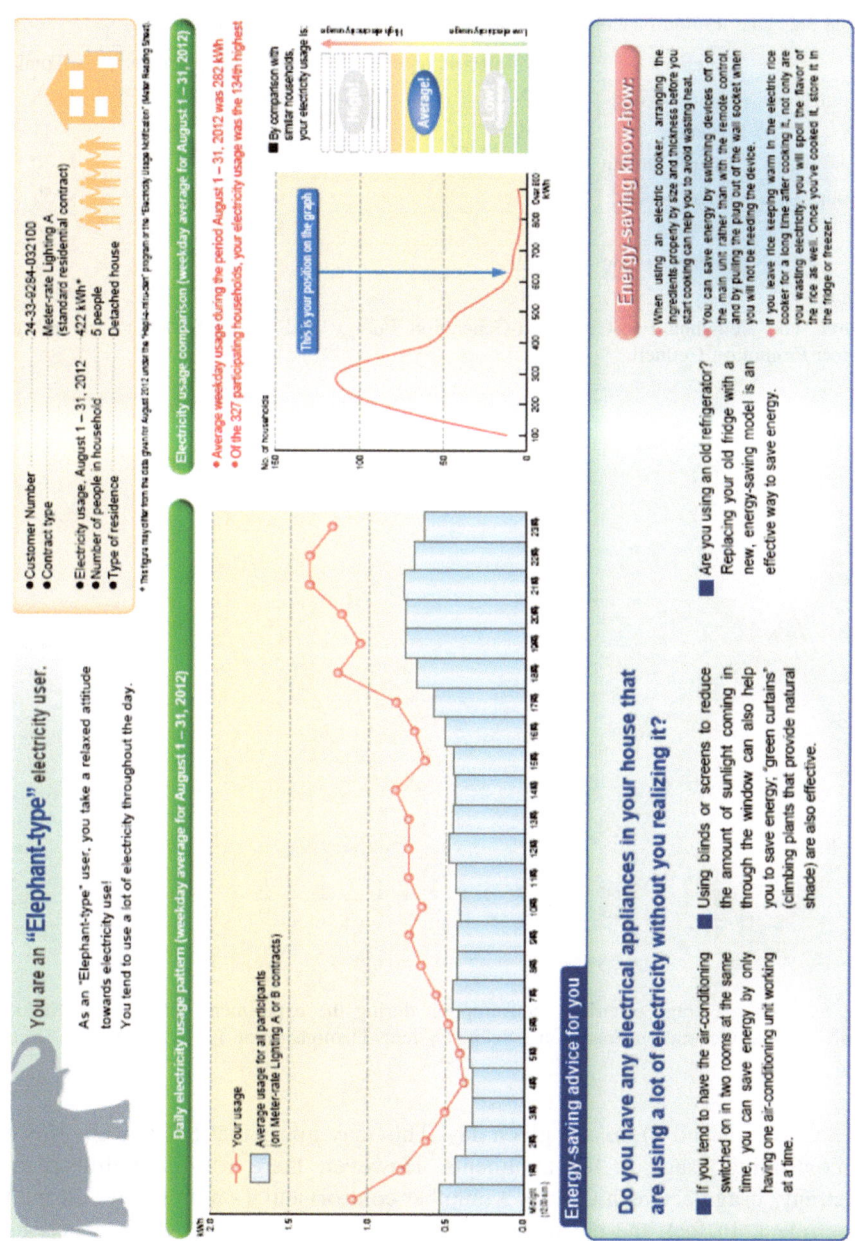

Fig. 5.2 Example of an HER. *Source* The Kansai Electric Power Company

each household automatically recorded the data. Figure 5.3 compares the daily average consumption of electricity during the experiment across four groups. Group 1 (a control group) exhibited higher usage of electricity than the treatments

Table 5.2 Distribution of electricity usage category by electricity contracts

	Category	All-electric contract	Standard contracts	Total
Group 2	Energy-using	13	25	38
	Average	18	23	41
	Energy-saving	16	24	40
Group 4	Energy-using	18	41	59
	Average	17	37	54
	Energy-saving	9	42	51
Total		91	192	283

Source The Keihanna Eco-City Next-Generation Energy and Social Systems Demonstration Project Promotion Council

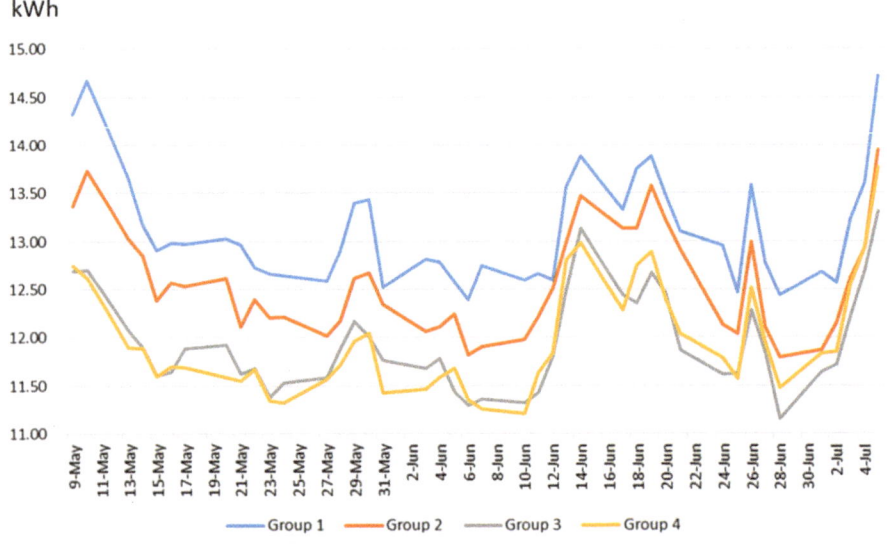

Fig. 5.3 Daily average electricity consumption during the experiment. *Source* The Keihanna Eco-City Next-Generation Energy and Social Systems Demonstration Project Promotion Council

(i.e., Groups 2 and 4) on any given day. This may imply that HERs were effective in controlling usage in the treatments. However, there was little difference in electricity usage between Groups 3 (another control) and 4 on most days during the experiment. In fact, the electricity usage of the latter groups, which experienced CPP in 2012, was persistently lower during the experiment than that of Groups 1 and 2, which did not experience CPP in 2012. This persistent difference in electricity usage may imply that households in Groups 3 and 4 were accustomed to saving electricity prior to the experiment on HERs.

5.3   Estimation Results

5.3.1   Econometric Model of Half-Hourly Electricity Consumption

A reduced-form model of household demand for electricity is used to estimate the effect of HERs on half-hourly electricity consumption. The focus of the analysis is on half-hourly usage of electricity because the conservation effects of HERs could differ across hours within a day, as shown in Sect. 5.3.3. Specifically, the natural log of half-hourly electricity consumption is assumed to be a linear function of a set of dummy variables for the treatment, temporal and cross-sectional fixed effects, and unobserved errors. This function is as follows:

$$\log KWH_{i,t} = [(\alpha_0 + \alpha_1 CO_i)U_i + (\alpha_2 + \alpha_3 CO_i)A_i + (\alpha_4 + \alpha_5 CO_i)S_i] \times \\ (\alpha_6 D_{2,i} + \alpha_7 D_{4,i}) \times HER_{i,t} + \omega_t + v_i + u_{i,t}, \tag{5.1}$$

where

$KWH_{i,t}$	half-hourly electricity consumption of household i on the tth time period, in kilowatt-hours (kWh),
CO_i	dummy variable equal to 1 if household i signs the all-electric contract,
U_i	dummy variable equal to 1 if household i is categorized as "energy-using,"
A_i	dummy variable equal to 1 if household i is categorized as "average,"
S_i	dummy variable equal to 1 if household i is categorized as "energy-saving,"
$D_{k,i}$	dummy variable equal to 1 if household i is allocated to Group k,
$HER_{i,t}$	dummy variable equal to 1 if household i received an HER before the t th time period,
ω_t	time fixed effects,
v_i	household fixed effects, and
$u_{i,t}$	error term.

The dummy variable $HER_{i,t}$ indicates whether the treatments received the report during the experiment. Due to the difference in the day when each household received a report as shown by Fig. 5.1, this dummy variable varies across both time and cross-section. Thus, the use of $HER_{i,t}$ identifies the impact of the report on electricity consumption in Eq. 5.1. For households in the control group, $HER_{i,t}$ becomes zero in any time period.

The conservation effects of HERs may depend on the neighbor comparison, electricity contracts, and experience of CPP. The impact of neighbor comparison is investigated by the use of three dummy variables associated with categories of electricity usage (i.e., U_i, A_i, and S_i). The dummy variable for the all-electric contract, (i.e., CO_i) which is assumed in Wolak (2011), is used in Eq. 5.1 to investigate the impact of electricity contracts. The experience of CPP, which suggests the persistent saving of electricity by the households receiving CPP, is indicated by the group dummy variables (i.e., $D_{2,i}$ and $D_{4,i}$). The parameters are indicated by $\alpha_0-\alpha_7$.

A two-way fixed effects model is assumed to account for temporal and cross-sectional effects on household electricity consumption. Household fixed effects include the impacts of income, appliance ownership, and demographic and dwelling characteristics; while time fixed effects include the impact of weather conditions and time-specific effects. Standard errors are clustered at the household level to correct for serial correlation in each household's electricity consumption. A two-way random effects model is also estimated for comparison. It is noteworthy that the Hausman statistic, which determines whether the estimation should be done using a random or fixed effects model, could not be computed because the difference of the covariance matrices failed to be positive definite (Greene 2007, E11–36).

5.3.2   Conservation Effects of HERs

Table 5.3 summarizes the estimation results of the two-way fixed and random effects models in Eq. 5.1. The data on weekends and holidays were excluded in the estimation of Eq. 5.1 because the neighbor comparison in the report was based on electricity usage on weekdays. The number of observations was the product of the number of weekdays during the experiment (i.e., 42 days), the number of half-hours a day (i.e., 48), and the number of households (i.e., 571). Table 5.4 summarizes the estimated coefficients of the HER dummy for households signing all-electric contracts.

The conservation effects of HERs depend on electricity contracts. As shown by Table 5.4, except for the "energy-saving" households in Group 2, the coefficients of the HER dummy were statistically significant for households with all-electric contracts in both the fixed and random effects models. In contrast, for both models, the coefficients on the HER dummy for households with standard contracts, which are shown in Table 5.3, were statistically insignificant even at the 10 % level.

The conservation effects for all-electric households range from 4.0 % to 8.7 %, depending on the category of electricity usage (i.e., U_i, A_i, or S_i) and the experience of CPP (i.e., $D_{2,i}$ or $D_{4,i}$). The "energy-using" category in the neighbor comparison, which was the highest tercile of the pre-treatment electricity consumption, had a larger impact on electricity saving than the other categories in Group 2 where households had no experience of receiving CPP prior to the HER experiment. The conservation effect of the "energy-using" category, which is 8.7 % and 8.0 % in the fixed and random effects models respectively, was larger than that in the previous literature. For instance, households in the highest decile of the pre-treatment usage were found to reduce electricity consumption by 6.3 % in Allcott (2011) and 7.0 % in Ayres et al. (2009). This difference implies that the conservation effects of HERs could be enhanced by the professional staff's visit to explain the energy-conservation tips. For the Group 4 households that experienced CPP prior to the

Table 5.3 The effects of HERs: estimation results of the two-way fixed and random effects models of half-hourly electricity consumption in Eq. 5.1

Variables	Fixed effects	Random effects
$D_{2,i} \cdot U_i \cdot HER_{i,t} \cdot CO_i$	$-0.101 \ (0.027)^{***}$	$-0.095 \ (0.027)^{***}$
$D_{2,i} \cdot A_i \cdot HER_{i,t} \cdot CO_i$	$-0.178 \ (0.129)$	$-0.174 \ (0.128)$
$D_{2,i} \cdot S_i \cdot HER_{i,t} \cdot CO_i$	$-0.137 \ (0.080)^{*}$	$-0.132 \ (0.080)^{*}$
$D_{4,i} \cdot U_i \cdot HER_{i,t} \cdot CO_i$	$-0.060 \ (0.023)^{***}$	$-0.056 \ (0.023)^{**}$
$D_{4,i} \cdot A_i \cdot HER_{i,t} \cdot CO_i$	$-0.045 \ (0.020)^{**}$	$-0.041 \ (0.020)^{**}$
$D_{4,i} \cdot S_i \cdot HER_{i,t} \cdot CO_i$	$-0.088 \ (0.026)^{***}$	$-0.085 \ (0.026)^{***}$
$D_{2,i} \cdot U_i \cdot HER_{i,t}$	$0.014 \ (0.024)$	$0.015 \ (0.024)$
$D_{2,i} \cdot A_i \cdot HER_{i,t}$	$0.131 \ (0.127)$	$0.130 \ (0.126)$
$D_{2,i} \cdot S_i \cdot HER_{i,t}$	$0.092 \ (0.075)$	$0.086 \ (0.075)$
$D_{4,i} \cdot U_i \cdot HER_{i,t}$	$0.015 \ (0.014)$	$0.016 \ (0.013)$
$D_{4,i} \cdot A_i \cdot HER_{i,t}$	$-0.017 \ (0.014)$	$-0.017 \ (0.013)$
$D_{4,i} \cdot S_i \cdot HER_{i,t}$	$0.027 \ (0.018)$	$0.026 \ (0.018)$
Adjusted R^2	0.3732	0.0004
Number of households	571	571
Number of observations	1,151,136	1,151,136

Notes The dependent variable is the natural log of half-hourly electricity consumption. The list of the explanatory variables is given in Eq. 5.1. Standard errors, which are clustered at the household level, are in parentheses
*Significant at the 10 % level
**Significant at the 5 % level
***Significant at the 1 % level

Table 5.4 The effects of HERs on electricity usage of all-electric households

Group	Category	Fixed effects	Random effects
Group 2	Energy-using	$-0.087 \ (0.015)^{***}$	$-0.080 \ (0.014)^{***}$
	Average	$-0.047 \ (0.026)^{*}$	$-0.044 \ (0.026)^{*}$
	Energy-saving	$-0.045 \ (0.028)$	$-0.045 \ (0.028)$
Group 4	Energy-using	$-0.046 \ (0.019)^{**}$	$-0.040 \ (0.019)^{**}$
	Average	$-0.061 \ (0.016)^{***}$	$-0.058 \ (0.016)^{***}$
	Energy-saving	$-0.061 \ (0.021)^{***}$	$-0.060 \ (0.021)^{***}$

Notes The effects and standard errors are computed from the estimation results in Table 5.3. Standard errors, which are clustered at the household level, are in parentheses
*Significant at the 10 % level
**Significant at the 5 % level
***Significant at the 1 % level

HER experiment, the "energy-using" category exhibited a lower conservation effect than the other categories; thus, households in the highest tercile of the pre-treatment usage reduced electricity consumption by 4.6 and 4.0 % in the fixed and random effects models, respectively. Overall, the "boomerang" effects, which cause households that used less than the social norm amount to use more, were not found.

As Fig. 5.3 illustrates, households in Groups 3 and 4 may have been accustomed to saving electricity prior to the experiment on HERs. To separate the effect of the electricity-saving habit from the conservation impact of HERs, the data on Groups 3 and 4 were used to estimate the model in Eq. 5.1 (see Table A.1 in the appendix). By comparison, the data on Groups 1 and 2 were used to estimate the model (see Table A.2 in the appendix). The estimated coefficients using the data on Groups 3 and 4 are close to those in Table 5.3. The latter is also true for the coefficients using the data on Groups 1 and 2. Thus, the estimated results of Eq. 5.1 do not depend on the effect of the electricity-saving habit, which, if any, is taken into account by either fixed or random effects.

5.3.3   Hourly Effects of HERs

The conservation effects of HERs may depend on household occupancy. At times, changes in behaviors that are associated with electricity usage are expected to occur when household members are present. In fact, using the data from the experiment of real-time electricity information feedback, Houde et al. (2013) found that electricity savings during the morning and evening peak periods were large and significant, while those during the middle of the day and night were insignificant.

The half-hourly treatment effects of HERs are estimated by the following electricity consumption equation:

$$\log KWH_{i,t} = \sum_{h=1}^{48} \left\{ \left[\beta_1^h U_i + \beta_2^h A_i + \beta_3^h S_i \right] \times (\beta_4^h D_{2,i} + \beta_5^h D_{4,i}) \times HOUR_t^h \right\}$$
$$\times HER_{i,t} + \omega_t + v_i + u_{i,t}, \tag{5.2}$$

where $HOUR_t^h$ takes a value of 1 during the hth time period and 0 otherwise for $h = 1$, ..., 48. In Eq. 5.2, the coefficient on each dummy variable (i.e., $\beta_1^h - \beta_5^h$) is assumed to differ across time of day. The half-hourly electricity consumption model in Eq. 5.2 is estimated only for households with all-electric contracts because the conservation effects were insignificant for households with standard contracts in Sect. 5.3.2.

Tables 5.5 and 5.6 present the estimated coefficients of the three dummy variables associated with the neighbor comparison during the morning and evening for Groups 2 and 4, respectively. These coefficients during the middle of the day and night, most of which were insignificant, are not listed in these tables. The standard errors are clustered at the household level to correct for serial correlation in the electricity consumption of each household. Tables 5.5 and 5.6 present the estimated coefficients in a two-way fixed effects model of Eq. 5.2. Similar results to those coefficients were obtained from estimating a two-way random effects model.

For the Group 2 households that had no experience of CPP, the "energy-using" category in the neighbor comparison had significant saving effects in the evening. These saving effects ranged from 12.1 to 20.2 %, which exceeded the average saving

Table 5.5 The effects of HERs on Group 2 during the morning and evening

Category	Hours	Coefficient
Energy-using	6:00–6:30 a.m.	−0.076 (0.151)
	6:30–7:00 a.m.	−0.093 (0.103)
	7:00–7:30 a.m.	−0.143 (0.089)
	7:30–8:00 a.m.	0.015 (0.069)
	8:00–8:30 a.m.	0.089 (0.074)
	8:30–9:00 a.m.	0.006 (0.065)
	6:00–6:30 p.m.	0.118 (0.077)
	6:30–7:00 p.m.	−0.051 (0.084)
	7:00–7:30 p.m.	−0.089 (0.102)
	7:30–8:00 p.m.	−0.179 (0.075)**
	8:00–8:30 p.m.	−0.202 (0.074)***
	8:30–9:00 p.m.	−0.147 (0.066)**
	9:00–9:30 p.m.	−0.121 (0.066)*
	9:30–10:00 p.m.	−0.071 (0.079)
Average	6:00–6:30 a.m.	−0.244 (0.139)*
	6:30–7:00 a.m.	−0.166 (0.120)
	7:00–7:30 a.m.	−0.104 (0.089)
	7:30–8:00 a.m.	0.072 (0.095)
	8:00–8:30 a.m.	0.063 (0.058)
	8:30–9:00 a.m.	0.025 (0.065)
	6:00–6:30 p.m.	0.164 (0.089)*
	6:30–7:00 p.m.	0.119 (0.075)
	7:00–7:30 p.m.	0.123 (0.069)*
	7:30–8:00 p.m.	0.142 (0.081)*
	8:00–8:30 p.m.	0.112 (0.066)*
	8:30–9:00 p.m.	0.069 (0.061)
	9:00–9:30 p.m.	0.088 (0.069)
	9:30–10:00 p.m.	0.018 (0.079)
Energy-saving	6:00–6:30 a.m.	0.026 (0.182)
	6:30–7:00 a.m.	−0.082 (0.141)
	7:00–7:30 a.m.	0.202 (0.102)**
	7:30–8:00 a.m.	0.112 (0.115)
	8:00–8:30 a.m.	0.025 (0.107)
	8:30–9:00 a.m.	−0.073 (0.077)
	6:00–6:30 p.m.	0.261 (0.118)**
	6:30–7:00 p.m.	0.152 (0.089)*
	7:00–7:30 p.m.	0.104 (0.087)
	7:30–8:00 p.m.	0.032 (0.113)
	8:00–8:30 p.m.	−0.004 (0.100)
	8:30–9:00 p.m.	0.042 (0.083)
	9:00–9:30 p.m.	−0.044 (0.130)
	9:30–10:00 p.m.	0.011 (0.103)

Notes The coefficients are obtained from estimating a two-way fixed effects model in Eq. 5.2 with data on households contracting all-electric. Standard errors, which are clustered at the household level, are in parentheses
*Significant at the 10 % level
**Significant at the 5 % level
***Significant at the 1 % level

Table 5.6 The effects of
HERs on Group 4 during the
morning and evening

Category	Hours	Coefficient
Energy-using	6:00–6:30 a.m.	−0.219 (0.129)*
	6:30–7:00 a.m.	−0.191 (0.115)*
	7:00–7:30 a.m.	−0.085 (0.094)
	7:30–8:00 a.m.	−0.050 (0.081)
	8:00–8:30 a.m.	−0.022 (0.070)
	8:30–9:00 a.m.	−0.001 (0.059)
	6:00–6:30 p.m.	0.046 (0.070)
	6:30–7:00 p.m.	−0.072 (0.066)
	7:00–7:30 p.m.	−0.104 (0.067)
	7:30–8:00 p.m.	−0.132 (0.073)*
	8:00–8:30 p.m.	−0.106 (0.064)*
	8:30–9:00 p.m.	−0.093 (0.064)
	9:00–9:30 p.m.	−0.125 (0.074)*
	9:30–10:00 p.m.	−0.155 (0.078)**
Average	6:00–6:30 a.m.	−0.328 (0.136)**
	6:30–7:00 a.m.	−0.245 (0.115)**
	7:00–7:30 a.m.	−0.035 (0.080)
	7:30–8:00 a.m.	0.021 (0.061)
	8:00–8:30 a.m.	0.060 (0.057)
	8:30–9:00 a.m.	0.012 (0.059)
	6:00–6:30 p.m.	−0.089 (0.081)
	6:30–7:00 p.m.	−0.136 (0.070)**
	7:00–7:30 p.m.	−0.128 (0.049)*
	7:30–8:00 p.m.	−0.061 (0.054)
	8:00–8:30 p.m.	−0.018 (0.067)
	8:30–9:00 p.m.	−0.050 (0.064)
	9:00–9:30 p.m.	−0.079 (0.061)
	9:30–10:00 p.m.	−0.098 (0.062)
Energy-saving	6:00–6:30 a.m.	0.156 (0.189)
	6:30–7:00 a.m.	−0.037 (0.167)
	7:00–7:30 a.m.	−0.070 (0.140)
	7:30–8:00 a.m.	−0.064 (0.092)
	8:00–8:30 a.m.	−0.148 (0.074)**
	8:30–9:00 a.m.	−0.119 (0.059)**
	6:00–6:30 p.m.	0.129 (0.113)
	6:30–7:00 p.m.	0.076 (0.090)
	7:00–7:30 p.m.	0.093 (0.133)
	7:30–8:00 p.m.	0.012 (0.103)
	8:00–8:30 p.m.	0.005 (0.089)
	8:30–9:00 p.m.	0.077 (0.100)
	9:00–9:30 p.m.	0.089 (0.094)
	9:30–10:00 p.m.	0.040 (0.093)

Notes The coefficients are obtained from estimating a two-way fixed
effects model in Eq. 5.2 with data on households contracting all-electric.
Standard errors, which are clustered at the household level, are in
parentheses
*Significant at the 10 % level
**Significant at the 5 % level
***Significant at the 1 % level

effect reported in Sect. 5.3.2. The saving effects of HERs were not found to be statistically significant in the morning for households labeled as "energy-using" in Group 2. The "energy-saving" and "average" categories of the pre-treatment electricity usage in Group 2 had adverse and significant effects on electricity saving in the evening. These adverse effects of peer comparison on electricity saving imply that the "boomerang" effects are likely to occur at times when household members are present.

Turning to the Group 4 households that experienced CPP prior to the HER experiment, all three categories of the pre-treatment electricity consumption in the neighbor comparison exhibited statistically significant conservation effects. Households labeled "energy-using" or "average," significantly reduced electricity consumption during both the morning and evening, while those labeled as "energy-saving" significantly reduced electricity consumption only during the morning. These saving effects ranged from 10.6 to 32.8 %, which exceeded the average savings reported in Sect. 5.3.2.

5.3.4 Persistence in the Conservation Effects of HERs

The persistence in the energy-conservation effect of a policy intervention affects its cost effectiveness, because long-lasting effects raise the benefits of interventions relative to the implementation costs. The previous studies examined the duration of conservation effects after households received the HERs. For example, Ayres et al. (2009) found that the reduction in electricity usage continued over 7–12 months after households began to receive the reports. Allcott and Rogers (2014) indicated that the decay rate of conservation effects was approximately 10–20 % per year for households that stopped receiving either monthly or quarterly reports after two years. This decay rate is four to eight times slower than the decay of conservation effects between the reports. The persistent conservation effects may be due to the accumulation of experience in saving electricity, which is close to the concept of "consumption capital" in Becker and Murphy (1988).

Due to the relatively short time period of the data, this section focuses on the persistence of conservation effects over the weeks after each household received the report. To test if the conservation effects of HERs persist over these weeks, a set of dummy variables indicating the time since each household received the report are added to the half-hourly electricity consumption model in Eq. 5.1. Specifically, the following half-hourly electricity consumption equation is estimated to examine the dependence of the coefficients on the time since the treatments received HERs:

$$\log KWH_{i,t} = \sum_{w=1}^{6} \left\{ \left[(\alpha_0^w + \alpha_1^w CO_i)U_i + (\alpha_2^w + \alpha_3^w CO_i)A_i + (\alpha_4^w + \alpha_5^w CO_i)S_i \right] \times (\alpha_6^w D_{2,i} + \alpha_7^w D_{4,i}) \times D_{i,t}^w \right\}$$
$$\times HER_{i,t} + \omega_t + v_i + u_{i,t},$$

$$(5.3)$$

where $D_{i,t}^w$ takes a value of 1 during the wth week of the treatment and 0 otherwise for $w = 1,\ldots, 5$. $D_{i,t}^6$ takes a value of 1 for all the weeks after the fifth week of the treatment and 0 otherwise. The number of dummy variables $D_{i,t}^w$ is assumed to be six, because the length of the period after receiving HERs is at most seven weeks.

Table 5.7 summarizes the estimated coefficients of three dummy variables associated with the neighbor comparison for households with all-electric contracts in Groups 2 and 4, respectively. The coefficients for households with standard contracts were insignificant; therefore, they are not listed in the table. The standard errors are clustered at the household level to correct for serial correlation in each household's electricity consumption. The coefficients were estimated by the two-way fixed effects model in Eq. 5.3. Similar results to those coefficients were obtained from estimating a two-way random effects model.

The results estimated in Table 5.7 imply that the effects of HERs on electricity usage depend on the time since households received the reports. They also imply that the persistence of the conservation effects over time differs across both categories of electricity consumption and groups. As an illustrative example, Fig. 5.4 compares the effects of HERs on electricity saving between households labeled as "energy-using" in Group 2 and those labeled as "average" in Group 4. Households labeled as "energy-using" in Group 2 reduced electricity usage after receiving the reports, and their reduction continued over several weeks. The conservation effect

Table 5.7 The weekly effects of HERs on electricity usage of all-electric households

Group	Category	Week	Coefficient, all-electric
Group 2	Energy-using	First	$-0.027 \ (0.026)$
		Second	$-0.058 \ (0.031)^*$
		Third	$-0.036 \ (0.028)$
		Fourth	$-0.035 \ (0.024)$
		Fifth	$-0.060 \ (0.016)^{***}$
		Sixth and after	$-0.071 \ (0.016)^{***}$
	Average	First	$-0.036 \ (0.029)$
		Second	$-0.021 \ (0.036)$
		Third	$-0.028 \ (0.031)$
		Fourth	$-0.037 \ (0.023)$
		Fifth	$-0.040 \ (0.256)$
		Sixth and after	$-0.035 \ (0.029)$
	Energy-saving	First	$-0.007 \ (0.019)$
		Second	$-0.009 \ (0.030)$
		Third	$-0.059 \ (0.040)$
		Fourth	$-0.051 \ (0.036)$
		Fifth	$-0.002 \ (0.039)$
		Sixth and after	$-0.007 \ (0.039)$

(continued)

Table 5.7   (continued)

Group	Category	Week	Coefficient, all-electric
Group 4	Energy-using	First	−0.009 (0.023)
		Second	−0.019 (0.021)
		Third	−0.007 (0.017)
		Fourth	−0.054 (0.021)**
		Fifth	−0.039 (0.018)**
		Sixth and after	−0.024 (0.027)
	Average	First	−0.030 (0.017)*
		Second	−0.070 (0.023)***
		Third	−0.058 (0.028)**
		Fourth	−0.064 (0.024)***
		Fifth	−0.062 (0.015)***
		Sixth and after	−0.031 (0.021)
	Energy-saving	First	0.009 (0.034)
		Second	−0.007 (0.041)
		Third	−0.021 (0.037)
		Fourth	−0.043 (0.047)
		Fifth	−0.072 (0.042)*
		Sixth and after	−0.028 (0.027)
Adjusted R^2			0.3732
Number of households			571
Number of observations			1,151,136

Notes The dependent variable is the natural log of half-hourly electricity consumption. The coefficients are obtained from estimating a two-way fixed effects model in Eq. 5.3 using data on all households. Standard errors, which are clustered at the household level, are in parentheses
*Significant at the 10 % level
**Significant at the 5 % level
***Significant at the 1 % level

on these households was statistically significant at the 1 % level for the fifth and sixth weeks after receiving the reports; these effects exceeded those for the first four weeks. In contrast, the conservation effects of HERs faded over time for households labeled as "average" in Group 4. Electricity saving by households labeled as "average" in Group 4 reached the maximum in the second week after receiving the reports, but declined afterwards.

The conservation effects of HERs also faded over time for households labeled as "energy-using" in Group 4. They reduced electricity usage mostly in the fourth week after receiving the reports, but their saving of electricity diminished in the fifth and sixth weeks. The conservation effects were insignificant in any given week for households labeled "energy-saving" or "average" in Group 2.

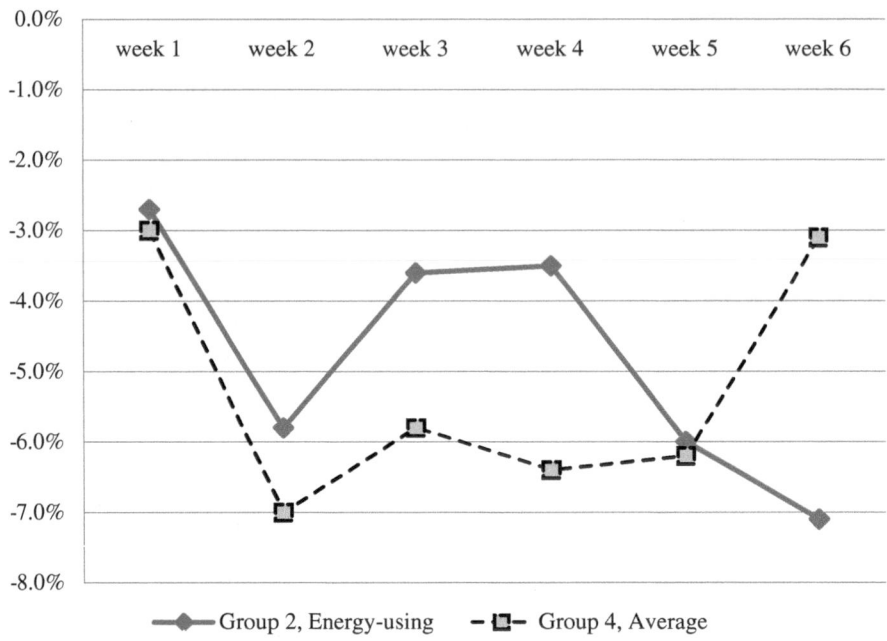

Fig. 5.4 Conservation effects over time: "energy-using" in Group 2 and "average" in Group 4

5.4   Conclusion

To examine the impact of HERs on electricity saving, this chapter conducts an econometric analysis of half-hourly electricity consumption of households using data from a randomized field experiment in 2013. In the experiment, households received personalized advice on the efficient use of energy appliances and peer comparisons of electricity usage when the experiment staff visited their homes. The peer comparisons indicated which of the three categories of electricity usage, that is, "energy-using," "average," and "energy-saving" was applied to each household on the basis of monthly electricity consumption prior to the experiment.

The estimated results of the half-hourly electricity consumption on weekdays during May 9 through July 5 indicate that HERs reduced electricity consumption of households with all-electric contracts but did not affect that of households with standard contracts. The electricity-saving effect of HERs on all-electric households, which ranged from 4.0–8.7 %, was found in each category of electricity usage. Overall, there was no indication of so-called "boomerang" effects, which raise electricity usage of households categorized as "energy-saving." The electricity-saving effect of HERs was found to become larger during the mornings and evenings of weekdays. This effect could be persistent over several weeks after providing HERs to some groups of households.

The dependence of HER effects on electricity contracts implies that the targeted provision of HERs would be an effective conservation policy (Allcott and Taubinsky 2015). The targeted provision of HERs is also effective in the long run, as implied by the dependence of the persistent effects of HERs on the neighbor comparisons. Future research could examine the group of households, which HERs should target, perhaps, by using the data on demographic characteristics and appliance holdings of households.

Appendix: Additional Estimation Results

See Tables A.1 and A.2.

Table A.1 The effects of HERs: estimation results of the two-way fixed and random effects models of half-hourly electricity consumption in Eq. 5.1, using data on Groups 3 and 4

Variables	Fixed effects	Random effects
$D_{4,i} \cdot U_i \cdot HER_{i,t} \cdot CO_i$	$-0.062\ (0.023)^{***}$	$-0.058\ (0.023)^{***}$
$D_{4,i} \cdot A_i \cdot HER_{i,t} \cdot CO_i$	$-0.045\ (0.019)^{**}$	$-0.042\ (0.019)^{**}$
$D_{4,i} \cdot S_i \cdot HER_{i,t} \cdot CO_i$	$-0.089\ (0.027)^{***}$	$-0.086\ (0.027)^{***}$
$D_{4,i} \cdot U_i \cdot HER_{i,t}$	$0.016\ (0.013)$	$0.017\ (0.013)$
$D_{4,i} \cdot A_i \cdot HER_{i,t}$	$-0.016\ (0.014)$	$-0.015\ (0.013)$
$D_{4,i} \cdot S_i \cdot HER_{i,t}$	$0.028\ (0.018)$	$0.027\ (0.018)$
Adjusted R^2	0.4402	0.0002
Number of households	323	323
Number of observations	651,168	651,168

Notes The dependent variable is the natural log of half-hourly electricity consumption. The list of the explanatory variables is given in Eq. 5.1. Standard errors, which are clustered at the household level, are in parentheses
**Significant at the 5 % level
***Significant at the 1 % level

Table A.2 The effects of HERs: estimation results of the two-way fixed and random effects models of half-hourly electricity consumption in Eq. 5.1, using data on Groups 1 and 2

Variables	Fixed effects	Random effects
$D_{2,i} \cdot U_i \cdot HER_{i,t} \cdot CO_i$	$-0.099\ (0.027)^{***}$	$-0.093\ (0.027)^{***}$
$D_{2,i} \cdot A_i \cdot HER_{i,t} \cdot CO_i$	$-0.179\ (0.129)$	$-0.174\ (0.128)$
$D_{2,i} \cdot S_i \cdot HER_{i,t} \cdot CO_i$	$-0.140\ (0.081)^{*}$	$-0.134\ (0.081)^{*}$
$D_{2,i} \cdot U_i \cdot HER_{i,t}$	$0.019\ (0.025)$	$0.017\ (0.025)$
$D_{2,i} \cdot A_i \cdot HER_{i,t}$	$0.131\ (0.127)$	$0.131\ (0.127)$

(continued)

Table A.2 (continued)

Variables	Fixed effects	Random effects
$D_{2,i} \cdot S_t \cdot HER_{i,t}$	0.091 (0.076)	0.087 (0.077)
Adjusted R^2	0.3365	0.0004
Number of households	248	248
Number of observations	499,968	499,968

Notes The dependent variable is the natural log of half-hourly electricity consumption. The list of the explanatory variables is given in Eq. 5.1. Standard errors, which are clustered at the household level, are in parentheses
*Significant at the 10 % level
***Significant at the 1 % level

References

Allcott H (2011) Social norms and energy conservation. J Public Econ 95:1082–1095

Allcott H (2015) Site selection bias in program evaluation. Quart J Econ 130:1117–1165

Allcott H, Kessler J (2015). The welfare effects of nudges: a case study of energy use social comparisons, NBER Working Paper 21671

Allcott H, Rogers T (2014) The short-run and long-run effects of behavioral interventions: experimental evidence from energy conservation. Am Econ Rev 104(10):3003–3037

Allcott H, Taubinsky D (2015) Evaluating behaviorally motivated policy: experimental evidence from the lightbulb market. Am Econ Rev 105(8):2501–2538

Arimura T, Katayama H, Sakudo M (2013) Do social norms matter to energy saving behavior? Endogenous social and correlated effects, Mimeo

Ayres I, Raseman S, Shih A (2009) Evidence from two large field experiments that peer comparison feedback can reduce residential energy usage, NBER Working Paper 15386

Becker G, Murphy K (1988) A theory of rational addiction. J Polit Econ 96:675–700

Costa D, Kahn M (2013) Energy conservation 'nudges' and environmentalist ideology: evidence from a randomized residential electricity field experiment. J Eur Econ Assoc 11(3):680–702

Ferraro P, Price M (2013) Using nonpecuniary strategies to influence behavior: evidence from a large-scale field experiment. Rev Econ Stat 95:64–73

Greene W (2007) LIMDEP version 9.0: econometric modeling guide, Vol. 1, Econometric Software Inc

Houde S, Todd A, Sudarshan A, Flora J, Armel C (2013) Real-time feedback and electricity consumption: a field experiment assessing the potential for savings and persistence. Energ J 34 (1):87–102

LaRiviere J, Holladay S, Novgorodsky D, Price M (2014) Prices vs. nudges: a large field experiment on energy efficiency fixed cost investments, Mimeo

Matsukawa I (2011) How does information provision affect residential energy conservation? Evidence from a field experiment. Energ Stud Rev 18(1):1–19

Nolan J, Schultz W, Cialdini R, Goldstein N, Griskevicius V (2008) Normative social influence is underdetected. Pers Soc Psychol Bull 34:913–923

Schultz W, Nolan J, Cialdini R, Goldstein N, Griskevicius V (2007) The constructive, destructive, and reconstructive power of social norms. Psychol Sci 18:429–434

Wolak F (2011) Do residential customers respond to hourly prices? Evidence from a dynamic pricing experiment. Am Econ Rev Pap Proc 101(3):83–87

Wooldridge J (2002) Econometric analysis of cross section and panel data, The MIT Press

Chapter 6
Regional Impacts of Energy Conservation Policies

Abstract This chapter assesses the impact of applying critical peak pricing (CPP), in-home displays (IHDs), and home energy reports (HERs) to consumers living in the Kansai region based on the empirical results of field experiments presented in the previous chapters. For CPP, households with standard and all-electric contracts in the Kansai region are assumed to have paid peak prices ranging from 65 to 105 U.S. cents/kWh (1 U.S. dollar = 100 yen) for 16 days in summer 2013 and 21 days in winter 2013/2014. On these days, these households are assumed to use IHDs once a day. The combination of CPP and IHDs would have reduced residential electricity usage during the peak period by approximately 15.0 % in summer 2013 and 19.7 % in winter 2013/2014. Overall, the effects of CPP together with IHD usage on the peak electricity demand (kilowatt) would be far larger than the effects of the government's call for a voluntary reduction. HERs are expected to reduce residential electricity consumption, thus, by applying HERs that include a peer comparison of electricity usage and personalized conservation tips to households contracting all-electric in the Kansai region in June 2013, a household would have saved 4.9–8.3 %. The cost effectiveness of HERs, which is defined as the implementation costs of HERs per electricity saving, would have been far larger in the region than that in the United States.

Keywords Electricity supply · Electricity saving · Supply cost · Cost effectiveness · CO_2 emissions · Outage costs

6.1 Introduction

The energy-conservation effects of critical peak pricing (CPP), in-home displays (IHDs), and home energy reports (HERs) described in the previous chapters imply that these policy interventions could yield substantial impacts if they were applied to consumers living in the Kansai region. This chapter assesses the regional impact of applying CPP, IHDs, and HERs to households in this region, using the empirical findings in the previous chapters.

© The Author(s) 2016
I. Matsukawa, *Consumer Energy Conservation Behavior After Fukushima*,
SpringerBriefs in Economics, DOI 10.1007/978-981-10-1097-2_6

The focus of CPP and IHDs is on the regional electricity demand during peak hours on critical peak days when it is likely to reach the supply capacity. The targeted periods for CPP are the summer of 2013 and the winter of 2013/2014, when there were serious concerns about the security of electricity supply in the Kansai region. In fact, the regional demand for electricity reached 95 % of the available supply on August 9, 2013. The application of CPP could have contributed to a secure electricity supply by reducing households' usage of electricity during peak hours on critical peak days.

The targeted households are assumed to use IHDs once each critical peak day. The energy-using effect of the use of IHDs found in Chap. 4 suggests that providing an IHD alone may have an adverse impact on energy conservation. If the policy target of providing an IHD is energy conservation, its provision to households must be accompanied by additional policy instruments that are effective for energy conservation. As argued by Price (2015), providing an IHD should be viewed as a complement to pecuniary strategies such as CPP. The empirical results of the field experiment in Chap. 4 imply that providing an IHD in combination with CPP is a promising option for conservation policy to be effective. This combination of policy instruments is effective in reducing electricity consumption during the peak period, as indicated by empirical evidence that the energy-saving effect of CPP together with the use of an IHD dominates the energy-using effect of providing an IHD in Table 4.9 in Chap. 4.

The regional impacts of CPP together with IHDs on the peak usage of electricity, supply costs, and consumers' utility are predicted on the basis of the effects of CPP and IHDs in Chap. 4. The assessment of the regional impacts of CPP also includes the way it contributes to outage prevention and reduction of CO_2 emissions. Because of the lack of data, it is hard to estimate the implementation costs of CPP and IHDs. Consequently, instead of assessing cost effectiveness, which is defined as the ratio of the implementation costs of CPP and IHDs to the reduction in peak electricity usage, this chapter assesses the cost savings associated with the regional supply of electricity using data on electricity supply costs, the maximum electricity demand, and the available generation capacity on critical peak days.

The focus of HERs is on all-electric household contracts in the Kansai region, which are targeted because their electricity savings in response to HERs were found to be statistically significant, as shown in Chap. 5. HERs are assumed to include a peer comparison of electricity usage along with energy-conservation tips in the form of a leaflet, and a professional is to visit each targeted household to provide explanations on the reports. The regional impacts of HERs on electricity usage are predicted on the basis of the conservation effects estimates in Chap. 5. The cost effectiveness of HERs, which is defined as the ratio of the costs of leaflets and hiring staff to electricity usage reduction, is compared with that in the U.S. studies. The impact of HERs also includes CO_2 emissions reduction, predicted on the basis of average emission factors in the region.

6.2 Methodology and Data

6.2.1 CPP and IHD Scenario

Table 6.1 summarizes a set of assumptions on the key items of the CPP and IHD scenario, which presumes that all households contracting standard A, B, and all-electric [i.e., 98 % of households in the Kansai region (FEPCJ 2013)] receive high prices of electricity for peak hours (i.e., 1 p.m.–4 p.m. in summer 2013 and 6 p.m.–9 p.m. in winter 2013/2014) on critical peak days. On these days, the targeted households are assumed to use IHDs once a day. The number of the targeted households is approximately 10,150,000 (FEPCJ 2013). Electricity prices are assumed to be 65, 85, or 105 cents/kWh respectively, depending on the regional electricity demand–supply ratios on critical peak days. Except for peak hours on these days, electricity prices are considered constant and identical to the average electricity price under standard contracts (i.e., 25 cents/kWh) and the number of critical peak days is the same as in the experiments conducted during summer 2013 and winter 2013/2014.

Tables 6.2 and 6.3 present the maximum regional demand for electricity, regional demand–supply ratios, and peak electricity prices on critical peak days in summer 2013 and winter 2013/2014. The maximum demand and demand–supply ratios are based on the actual data. The highest price of electricity, 105 cents/kWh, is applied to households if the demand–supply ratio exceeds 90 % in the summer of 2013 and 88 % in the winter of 2013/2014. The price of 85 cents/kWh is applied if the ratio ranges from 87 to 90 % for summer 2013 and from 81 to 88 % for winter 2013/2014, while the price of 65 cents/kWh is applied for a range of less than 87 % in summer 2013 and 81 % in winter 2013/2014.

Regional electricity savings associated with CPP and IHDs are predicted on the basis of residential electricity demand in the Kansai region during peak hours and the estimates of the effects of CPP and IHDs in Chap. 4. Specifically, given the electricity price, electricity savings on each critical peak day are defined as the product of the daily electricity usage of households during peak hours in the region and the percentage reduction in that usage. The daily electricity usage of households during peak hours in the region was approximately 24 million kWh in summer 2013, and 30 million kWh in winter 2013/2014. The percentage reduction

Table 6.1 Assumptions on the key items of the CPP and IHD scenario

Items	Content
Targeted customers	CPP is applied to households contracting standard A, B, and all-electric (10,150,000 households)
IHD use	Each household uses an IHD once a day when CPP is applied
Electricity prices	Prices are 65, 85, and 105 cents/kWh
Number of critical peak days	CPP is applied to 16 days in summer and 21 days in winter

Table 6.2 Maximum regional demand for electricity, regional demand–supply ratio, and electricity prices on critical peak days in summer 2013

Date	Maximum regional demand (10,000 kW)	Demand–supply ratio (%)	Electricity prices (cents/kWh)
July 9	2,587	86	65
July 10	2,582	88	85
July 16	2,372	86	65
July 19	2,311	86	65
July 22	2,506	87	85
July 25	2,592	92	105
July 26	2,564	89	85
August 6	2,573	87	85
August 8	2,696	92	105
August 9	2,729	95	105
August 23	2,733	91	105
August 26	2,041	74	65
August 29	2,362	89	85
August 30	2,545	93	105
September 11	2,252	89	85
September 12	2,363	92	105

Note The data on the maximum regional demand and demand–supply ratios are obtained from the Kansai Electric Power Company

in daily electricity usage of households during these hours is obtained from the effects of CPP together with IHD usage on the peak electricity demand of households using IHD once a day in Table 4.9 in Chap. 4. Because of the lack of the estimates, these effects of summer 2012 and winter 2012/2013 are used to compute the percentage reduction in the electricity usage in summer 2013 and winter 2013/2014. For instance, in the summer, on critical peak days with prices of 105 cents/kWh, the percentage reduction in daily electricity usage is 15.0 % (i.e., 14.4 + 1.2 − 0.6) for households using IHDs once a day.

Given the duration (hours) of electric appliance usage during peak hours, the percentage reduction in residential electricity usage (kWh) indicates the percentage reduction in the residential maximum demand (kW) for electricity during peak hours. The product of the percentage reduction in the maximum demand associated with the price of 105 cents/kWh and the residential maximum demand during peak hours in the Kansai region (8 million kW in summer 2013 and 10 million kW in winter 2013/2014) yields the predicted reduction in the maximum demand for electricity in the region.

Supply cost savings associated with CPP and IHDs are defined as either avoided costs of operating oil-fired generation plants during peak hours or of procuring electricity from a wholesale market. For the former, the marginal cost of these

Table 6.3 Maximum regional demand for electricity, regional demand–supply ratio, and electricity prices on critical peak days in winter 2013/2014

Date	Maximum regional demand (10,000 kW)	Demand–supply ratio (%)	Electricity prices (cents/kWh)
December 6	2,016	79	65
December 11	2,149	83	85
December 12	2,254	86	85
December 16	2,253	89	105
December 17	2,215	83	85
December 18	2,289	86	85
December 26	2,186	82	85
December 27	2,124	79	65
January 7	2,270	83	85
January 10	2,344	88	85
January 15	2,309	82	85
January 17	2,290	85	85
January 21	2,347	89	105
January 22	2,374	89	105
January 23	2,329	83	85
January 27	2,277	80	65
January 29	2,136	77	65
February 5	2,392	84	85
February 6	2,451	89	105
February 10	2,360	90	105
February 13	2,370	90	105

Note The data on the maximum regional demand and demand–supply ratios are obtained from the Kansai Electric Power Company

plants of 37.6 cents/kWh, which is obtained from the Kansai Electric Power Company, is used to compute supply cost savings per reduced kWh. For the latter, the maximum wholesale price of electricity during peak hours (JEPX 2014) is used to compute supply cost savings per reduced kWh. The maximum wholesale price during peak hours was 55.0 cents/kWh in summer 2013 and 34.2 cents/kWh in winter 2013/2014.

6.2.2 HER Scenario

Table 6.4 summarizes the set of assumptions on the key items of the HERs scenario. HERs are assumed to consist of energy conservation tips and neighbor comparison of electricity usage. They target households contracting all-electric in the Kansai region, because, during the experiment, they were found to have statistically significant conservation effects on these households. The number of the

Table 6.4 Assumptions on the key items of the HER scenario

Items	Content
Targeted customers	HERs are applied to households contracting all-electric (1,155,000 households)
Peer comparison categories	Participating households are allocated to energy-saving, average, or energy-using
CPP experience	All participants experienced CPP

targeted households is approximately 1,155,000 (FEPCJ 2013). All-electric households in the region are allocated to the energy-saving, average, or energy-using group, depending on their historical usage of electricity and are assumed to have already experienced CPP prior to receiving HERs.

Participating households received HERs at the beginning of June 2013. Electricity savings in June 2013, predicted for the Kansai region, are computed by each category of peer comparison on the basis of the estimated impacts of HERs on electricity usage in Chap. 5. Specifically, the conservation effects of three categories of neighbor comparisons for all-electric households in Group 4, which range from 4.6–6.1 % in the fixed effects model, are used to compute the electricity savings. The implementation of the HER program is assumed to require leaflets and hiring of professional staff costs. The cost of a leaflet is assumed to be 1 dollar (Allcott and Rogers 2014), while that of hiring a professional is assumed to be 34 dollars/h, which is obtained from the ratio of the monthly average wage to the monthly average working hours in utilities (JMHLW 2014). A professional is assumed to visit each household for an hour to explain HERs.

6.3   Regional Impacts of CPP, IHDs, and HERs

Tables 6.5 and 6.6 summarize the impacts on electricity savings, supply cost savings, and CO_2 cost savings in the Kansai region under the CPP and IHD scenario for summer 2013 and winter 2013/2014, respectively. Under the CPP and IHD scenario, which targeted approximately 10,150,000 households contracting standard and all-electric (FEPCJ 2013), total electricity savings would be approximately 49.7 million kWh in the region or 4.9 kWh per household in summer 2013 and 103.0 million kWh in the region or 10.1 kWh per household in winter 2013/2014.

These electricity savings indicate that because of the highest price of 105 cents/kWh, a reduction in the residential peak demand for electricity would be approximately 1,200,000 kW or 15.0 % in the Kansai region on August 23, the day when the regional demand became the highest among the 16 critical peak days of summer 2013. They also indicate that the reduction would be approximately 1,970,000 kW or 19.7 % on February 6, the day when the regional demand became the highest among the 21 critical peak days of winter 2013/2014. The effects of CPP

Table 6.5 Regional impacts under the CPP and IHD scenario in summer 2013

Alternative electricity supply	Oil-fired generation	Wholesale purchase
Total electricity savings during peak hours in the region (million kWh)	49.7	
Peak electricity savings per household (kWh/household)	4.9	
Total reduction in the residential maximum demand for electricity (10,000 kW)	120.0	
Percentage reduction in the residential maximum demand for electricity (%)	15.0	
Total savings of electricity supply costs associated with peak usage in the region (million US$)	18.7	27.3
Supply cost savings per reduced electricity (US cents/kWh)	37.6	55.0
Total savings of the cost of CO_2 emissions associated with electricity usage during peak hours in the region (million US$)	1.2	
CO_2 cost savings per reduced electricity (US cents/kWh)	2.4	

Notes The CPP and IHD scenario assumes that approximately 98 % of households in the Kansai region would have received CPP during 1 p.m.–4 p.m. on 16 days in summer 2013. On these days, these households are assumed to use IHDs once a day. The number of the targeted households is approximately 10,150,000. The regional impacts of CPP and IHDs are computed from these effects on the residential peak demand for electricity in Chap. 4. The regional impacts on the costs of electricity supply and CO_2 emissions do not include the effects of CPP together with IHDs on the residential off-peak usage of electricity

together with IHD use on the peak electricity demand in Tables 6.5 and 6.6 are far larger than the effects of the government's call for a voluntary reduction in the peak demand for electricity. This voluntary reduction is found to reduce the maximum demand for residential electricity in the Kansai region by 9 % in summer 2013 (JMETI 2013, p. 15) and 5 % in winter 2013/14 (JMETI 2014, p. 12).

The combination of CPP and IHDs would have reduced the cost of supplying electricity to the entire region by 18.7 million dollars or 37.6 cents/kWh if oil-fired generation plants are used to supply electricity during peak hours on critical peak days in summer 2013 and 38.7 million dollars or 37.6 cents/kWh in winter 2013/2014. Reducing electricity consumption by 1 kWh during a peak period that lasts for three hours would be worth 37.6 cents under the CPP and IHD scenario. The cost saving would be 27.3 million dollars or 55.0 cents/kWh if the utility procures electricity from the wholesale market, Japan Electric Power Exchange, in order to supply electricity to the region during peak hours on critical peak days in summer 2013 and 35.2 million dollars or 34.2 cents/kWh in winter 2013/2014.

Although the focus of CPP is on the reduction of electricity usage during peak hours on critical peak days, it also contributes to a reduction in CO_2 emissions if oil-fired power plants are used to supply electricity during peak hours on critical peak days. We assume that the cost of CO_2 emissions amounts to 34.15 dollars/ton of CO_2, which is also assumed by the Tokyo Tax Commission (Arimura and Iwata

Table 6.6 Regional impacts under the CPP and IHD scenario in winter 2013/2014

Alternative electricity supply	Oil-fired generation	Wholesale purchase
Total electricity savings during peak hours in the region (million kWh)	103.0	
Peak electricity savings per household (kWh/household)	10.1	
Total reduction in the residential maximum demand for electricity (10,000 kW)	197.0	
Percentage reduction in the residential maximum demand for electricity (%)	19.7	
Total savings of electricity supply costs associated with peak usage in the region (million US$)	38.7	35.2
Supply cost savings per reduced electricity (US cents/kWh)	37.6	34.2
Total savings of the cost of CO_2 emissions associated with electricity usage during peak hours in the region (million US$)	2.4	
CO_2 cost savings per reduced electricity (US cents/kWh)	2.4	

Notes The CPP and IHD scenario assumes that approximately 98 % of households in the Kansai region would have received CPP during 6 p.m.–9 p.m. on 21 days in winter 2013/2014. On these days, these households are assumed to use IHDs once a day. The number of the targeted households is approximately 10,150,000. The regional impacts of CPP and IHDs are computed from these effects on the residential peak demand for electricity in Chap. 4. The regional impacts on the costs of electricity supply and CO_2 emissions do not include the effects of CPP and IHDs on the residential off-peak usage of electricity

2015, p. 154), and that the oil-fired electricity generation emits 0.691 kilograms of CO_2 per kWh (Uchiyama and Yamamoto 1992, p. 13). Under the CPP and IHD scenario, given the off-peak usage of electricity, the cost savings associated with CO_2 emissions during peak hours in the Kansai region would be approximately 1.2 million dollars in summer 2013 and 2.4 million dollars in winter 2013/2014. Thus, the CO_2 cost savings are much lower than the cost savings of electricity supply under the CPP and IHD scenario.

CPP may contribute to preventing outages by reducing the regional electricity demand on critical peak days, because the possibility that an outage occurs increases on these days, when the regional demand for electricity is close to available supply. The literature measures the economic costs of outage (Caves et al. 1992; Matsukawa and Fujii 1994). Suppose, for instance, that in the case of an outage during peak hours in August, the regional economy will incur an economic loss of approximately 4.87 dollars per interrupted electricity (kWh), which is given by the gross domestic product per electricity usage in 2013. The probability that outage occurs is given by the loss-of-load probability on a critical peak day in August, which is assumed to be 0.3 days in August (ESCJ 2009, 1–45), or 0.00968 (i.e., 0.3 ÷ 31). Then, the expected loss of the regional economy for the peak period of 3 hours would be 3,860,000 or 0.0471 dollars per interrupted kWh. This expected loss is larger than the total cost savings per day in summer 2013.

Table 6.7 summarizes the regional impacts on electricity savings and cost effectiveness for each category of electricity usage neighbor comparison under the HER scenario. Each targeted household contracting all-electric would reduce electricity consumption in June 2013 by 26.9–45.1 kWh, or 4.9–8.3 % relative to the average all-electric household that used 546.6 kWh in June 2012. This reduction depends on the neighbor comparison category. Monthly electricity savings would range from 10.3 to 17.4 million kWh in the region. Under the HER scenario, cost effectiveness, which is defined as the implementation costs of HERs per electricity savings, would range from 0.77 to 1.30 dollars/kWh. These estimates on cost effectiveness of HERs are much larger than the U.S. estimates. For instance, the cost effectiveness of providing a report to households is approximately 4 cents/kWh (Allcott and Rogers 2014). This estimate is similar to the average cost effectiveness across energy conservation programs in the United States, that is, approximately 5 cents/kWh (Arimura et al. 2011). The difference in the cost effectiveness of HERs between the estimates in this chapter and the U.S. estimates reflects the cost of hiring a professional. Consequently, HER cost effectiveness is smaller than U.S. estimates if the cost of hiring a professional is excluded, as shown by the column labeled "Leaflet" in Table 6.7.

Table 6.7 Regional impacts under the HER scenario in June 2013

Program cost items		Leaflet	Leaflet + staff
Monthly savings of residential electricity usage in the region (million kWh)	Energy-saving	10.3	
	Average	14.9	
	Energy-using	17.4	
Monthly electricity savings per household (kWh/household)	Energy-saving	26.9	
	Average	38.6	
	Energy-using	45.1	
Cost effectiveness of HERs (US$/kWh)	Energy-saving	0.04	1.30
	Average	0.03	0.91
	Energy-using	0.02	0.77
Monthly savings of CO_2 emissions associated with residential electricity usage in the region (thousand ton of CO_2)	Energy-saving	5.3	
	Average	7.7	
	Energy-using	9.0	
Monthly savings of the cost of CO_2 emissions associated with residential electricity usage in the region (million US$)	Energy-saving	0.18	
	Average	0.26	
	Energy-using	0.31	
CO_2 cost savings per reduced electricity (US cents/kWh)		1.8	

Notes The HER scenario assumes that a professional would have visited each household contracting all-electric in the Kansai region to explain HERs for an hour at the beginning of July in 2013. The number of the targeted households is approximately 1,155,000. The regional impacts are estimated from the conservation effects of HERs on half-hourly electricity consumption of all-electric households that experienced CPP in Chap. 5. The conservation effects of HERs are assumed to continue for a month. The cost effectiveness of HERs is defined as the ratio of the implementation costs to electricity savings

Energy conservation through providing HERs to households will also contribute to CO_2 emissions reduction. With the assumption that targeted all-electric households receiving HERs reduce electricity supplied by the utility whose average emission factors are 0.516 CO_2 kilograms/kWh in the region (KEPCO 2016), total savings of CO_2 emissions would amount to 22,000 ton of CO_2 in June 2013. Further, with the assumption that the cost of CO_2 emissions amounts to 34.15 dollars/ton of CO_2, cost savings would amount to approximately 1.8 cents/kWh in June 2013 and are much more modest compared to HER cost effectiveness when both leaflet and staff costs are taken into account. Total cost savings associated with CO_2 emissions in the Kansai region would be approximately 750,000 dollars in June 2013 under the HER scenario.

6.4 Conclusion

This chapter assesses the impact of applying CPP, IHDs, and HERs to consumers living in the Kansai region based on the empirical results of field experiments presented in the previous chapters. The CPP and IHD scenario is assumed to target approximately 10,150,000 households contracting standard and all-electric, and to apply electricity prices ranging from 65 to 105 cents/kWh to all targeted households in the Kansai region for 16 days in summer 2013 and 21 days in winter 2013/2014. On these days, these households are assumed to use IHDs once a day. The combination of CPP and IHDs would have saved residential electricity usage by approximately 49.7 million kWh in the region in summer 2013 and 103.0 million kWh in the region in winter 2013/2014. Overall, the effects of CPP together with IHD use on the peak electricity demand (kW) would be far larger than the effects of the government's call for voluntary reduction in the peak demand for electricity. The combination of CPP and IHDs would have saved costs of oil-fired generation plants by 37.6 cents/kWh. Alternatively, it would have saved costs of procuring electricity at a wholesale market by approximately 55.0 cents/kWh in summer 2013 and 34.2 cents/kWh in winter 2013/2014, respectively.

The impact of CPP on the region extends to the possible prevention of outages during peak hours on critical peak days. It may also contribute to CO_2 emission reductions associated with the operation of oil-fired generation plants. However, these benefits of CPP may be countervailed by the loss of household utility caused by high electricity prices and the implementation costs of CPP and IHDs, which are not taken into account in this chapter.

The HER scenario assumes that the staff would have explained the report consisting of energy conservation tips and neighbor comparisons of electricity usage to households contracting all-electric in the Kansai region at the beginning of June 2013. The targeted households are categorized as energy-saving, average, or energy-using, depending on their historical usage of electricity. HERs would have reduced monthly electricity consumption by 26.9–45.1 kWh per all-electric household in the region during June 2013. The conservation effect of HERs on monthly electricity consumption ranges from 4.9 to 8.3 % relative to the monthly electricity usage of the average all-electric household in the region. The cost effectiveness of HERs, which is defined as the implementation costs of HERs per electricity saving, would range from 0.77 to 1.30 dollars/kWh.

This chapter focuses only on the short-run impact of innovative instruments for energy conservation. The effectiveness of policy interventions depends on the persistence of conservation effects in the long run (Allcott and Rogers 2014). The impacts of CPP, IHDs, and HERs are persistent if households continue to respond to these interventions over years. Indeed, 91 % households that saved electricity during the summer of 2013 were willing to continue saving electricity during the summer of 2014 in the Kansai region (JMETI 2014, p. 36). This persistent behavior of energy conservation, if any, would favor the long-run effectiveness of policy interventions.

This chapter assumes the mandatory application of CPP to all households contracting standard and all-electric. In practice, CPP may be applied to households on a voluntary basis. If that is the case, the long-run impact of CPP depends on households' choice of CPP. A key issue of voluntary application of CPP is what factors influence households' choice of CPP. These factors have been investigated by the literature on time-of-use pricing of electricity (Caves et al. 1989; Train and Mehrez 1994; Baladi et al. 1998; Matsukawa 2001). Future research could apply the choice model of electricity prices as well as that of energy appliances to investigate the long-run impact of innovative instruments for energy conservation.

References

Allcott H, Rogers T (2014) The short-run and long-run effects of behavioral interventions: experimental evidence from energy conservation. Am Econ Rev 104(10):3003–3037

Arimura T, Iwata K (2015) An evaluation of Japanese environmental regulations, Springer

Arimura T, Li S, Newell R, Palmer K (2011) Cost-effectiveness of electricity energy efficiency programs, NBER Working Paper 17556

Baladi M, Herriges J, Sweeney T (1998) Residential response to voluntary time-of-use electricity rates. Resour Energy Econ 20:225–244

Caves D, Herriges J, Kuester K (1989) Load shifting under voluntary residential time-of-use rates. Energy J 10:83–99

Caves D, Herriges J, Windle R (1992) The cost of electric power interruptions in the industrial sector: estimates derived from interruptible service programs. Land Econ 68:49–61

ESCJ [Electric Power System Council of Japan] (2009) A survey on interconnectors in Japan (in Japanese)

FEPCJ [Federation of Electric Power Companies of Japan] (2013) Handbook of electric power industry, Ohmusha (in Japanese)

JEPX [Japan Electric Power Exchange] (2014) Spot price index, available at http://www.jepx.org/pdf/market/Index/Index_2014 (in Japanese). Accessed March 2015

JMETI [Japan Ministry of Economy, Trade and Industry] (2013) Report of a sub-committee on the review of electricity demand and supply, October (in Japanese)

JMETI [Japan Ministry of Economy, Trade and Industry] (2014) Report of a sub-committee on the review of electricity demand and supply, April (in Japanese)

JMHLW [Japan Ministry of Health, Labour and Welfare] (2014) Monthly labor survey
KEPCO (2016) Website of the Kansai Electric Power Company, http://www.kepco.co.jp/
    sustainability/kankyou/co2/index.html#co2, (in Japanese). Accessed January 2016
Matsukawa I (2001) Household response to optional peak-load pricing of electricity. J Regul Econ
    20(3):249–267
Matsukawa I, Fujii Y (1994) Customer preferences for reliable power supply: using data on actual
    choices of back-up equipment. Rev Econ Stat 76:434–446
Price M (2015) Using field experiments to address environmental externalities and resource
    scarcity: major lessons learned and new directions for future research, NBER Working Paper
    20870
Train K, Mehrez G (1994) Optional time-of-use prices for electricity: econometric analysis of
    surplus and Pareto impacts. RAND J Econ 25:263–283
Uchiyama Y, Yamamoto H (1992) Greenhouse effect analysis of power generation plants, Central
    Research Institute of Electric Power Industry Report Y91005 (in Japanese)